Malak Ben Romdhane
Mounir Frija

Study of the mechanical strength of implantable hip prostheses

Malak Ben Romdhane
Mounir Frija

Study of the mechanical strength of implantable hip prostheses

PHT in Ti-6Al-4V titanium superalloy

Imprint
Any brand names and product names mentioned in this book are subject to trademark, brand or patent protection and are trademarks or registered trademarks of their respective holders. The use of brand names, product names, common names, trade names, product descriptions etc. even without a particular marking in this work is in no way to be construed to mean that such names may be regarded as unrestricted in respect of trademark and brand protection legislation and could thus be used by anyone.

Cover image: www.ingimage.com

This book is a translation from the original published under ISBN 978-620-6-72629-6.

Publisher:
Sciencia Scripts
is a trademark of
Dodo Books Indian Ocean Ltd. and OmniScriptum S.R.L publishing group

120 High Road, East Finchley, London, N2 9ED, United Kingdom
Str. Armeneasca 28/1, office 1, Chisinau MD-2012, Republic of Moldova, Europe
Printed at: see last page
ISBN: 978-620-8-28538-8

TABLE OF CONTENTS

GENERAL INTRODUCTION

Many people suffer from severe pain due to deterioration of the hip joint, which is why they need hip prostheses to relieve pain and restore the joint's proper function.

Over the years, researchers and surgeons have worked side by side to improve the performance of prostheses and ensure the success of orthopaedic surgery. But sometimes the choice of materials used affects how well they work. That's why biomaterials are used to better adapt to the human body.

And despite remarkable advances in materials, design and manufacturing techniques, service life remains a research objective, as do problems of stem fracture and wear of the acetabulum and femoral head.

The femoral stem undergoes polycyclic stress during the patient's daily activities, which causes the possibility of breakage.

This book explores the origins of femoral stem cracking and fracture through the use of numerical modeling.

The numerical method is a means of experimenting with the model virtually. This method complements the experimental method for analyzing movements and deformations, especially when the geometric shapes of these bodies are complicated. The deformations they undergo are large, the materials they are made of have non-linear behavior and the loads applied are dynamic.

The book is divided into four chapters:

The first chapter is essentially devoted to the description of the hip prosthesis, illustrating a number of bibliographical studies on the history, composition, stress, fatigue and cracking of implantable stems, and brings together a number of studies relating to the different materials used and the methods of manufacturing hip prostheses.

The second chapter illustrates some existing research on fatigue testing and simulation mechanisms for hip prostheses.

The third chapter presents a simulation and numerical analysis using the finite element method and Ansys software.

The fourth chapter presents a simulation study of the polycyclic fatigue behavior of a femoral stem using nCode DesignLife software.

The book closes with a general conclusion and outlook.

CHAPITRE 1: BIBLIOGRAPHIC RESEARCH ON HIP PROSTHESES

1.1 Introduction

This chapter presents a literature search on coxofemoral prostheses, also known as hip prostheses. And to better understand how a hip prosthesis works, descriptive studies of the hip joint are illustrated, encompassing its bio-structure and possible movements. Synthetic studies of hip prostheses were presented, including their history, compositions, stresses and fatigue testing of implantable stems.

1.2 Hip anatomy

The hip joint, also known as the coxofemoral joint (Figure 1-1), is responsible for moving the lower limbs in all possible directions.

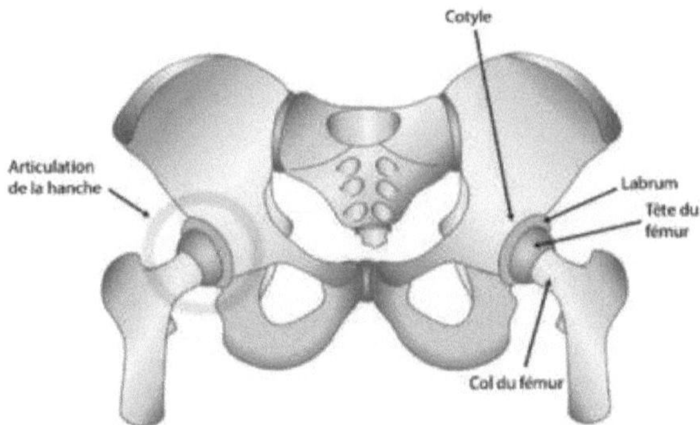

Figure 1- 1: Coxo-femoral joint

This joint is of the spheroid type (a joint whose surface forms a sphere, as shown in figure 1-2), with three axes and three degrees of freedom: the antero-posterior axis, which enables abduction and adduction movements; the transverse axis, which enables flexion and extension movements; and the longitudinal axis, which enables internal and external rotation movements (figure 1-3).[1][2] The joint is also a spheroid.

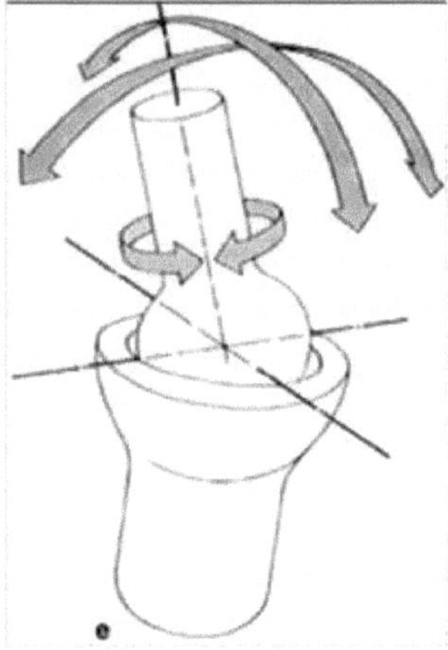

Figure 1- 2: The spheroid surface of the joint

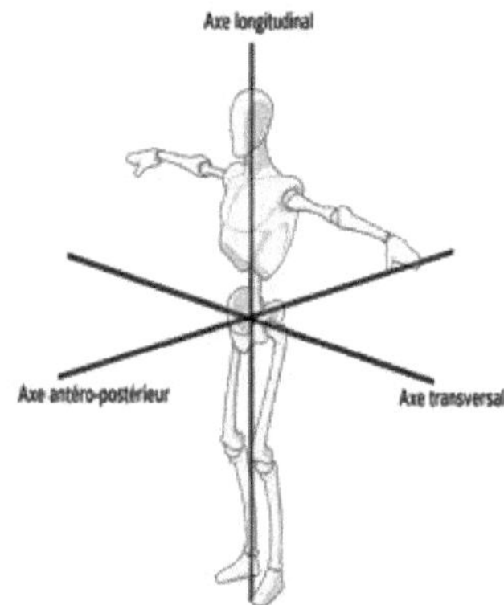

Figure 1- 3: The 3 axes that ensure degrees of freedom

1.2.1 Bone anatomy

The hip (figure 1-4) consists essentially of a pelvic bone, a femoral head that articulates in a hemispherical cavity in the pelvis called the acetabulum, whose contact surface is covered by cartilage, a femoral neck and a femur :

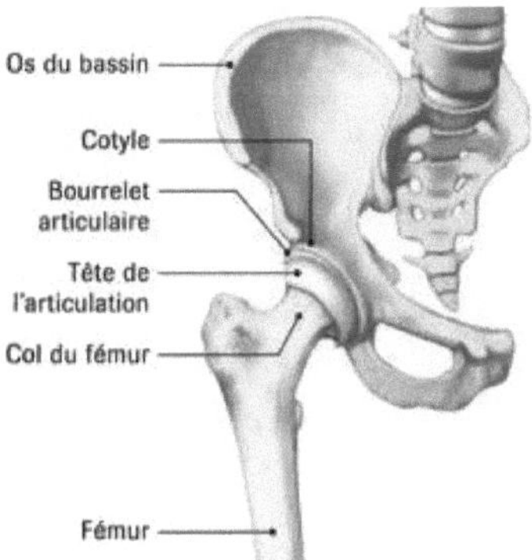

Figure 1- 4: Hip bone architecture

- Coxal bone: also known as the iliac bone, this is the main hip bone and connects the spine to the lower limbs. The two coxal bones form the pelvis or pelvic girdle (figure 1-5). The pelvic girdle is very stable, as it must withstand the loads it receives.

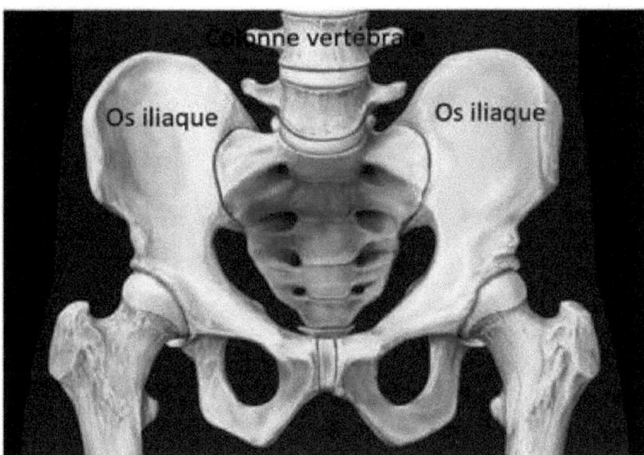

Figure 1- 5: The pelvic girdle

- The femur: also known as the thigh bone (figure 1-6), ends in the femoral head, which is shaped like a sphere. The femoral head nests with the iliac bone in a cavity called the acetabulum.

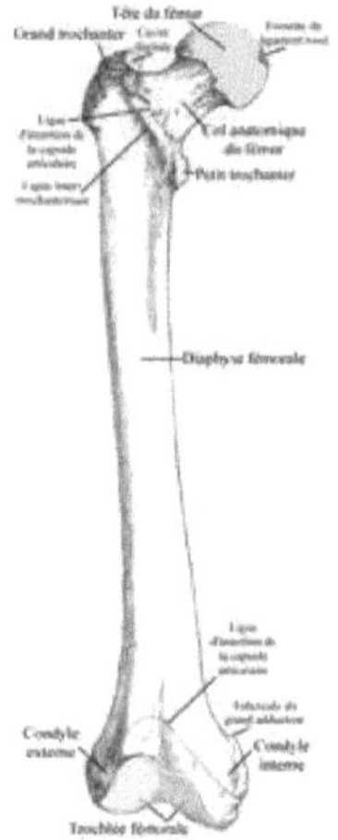

Figure 1- 6: Thigh bones

The ilium and femur are joined by an articular capsule and coxofemoral ligaments (Figure 1-7).

The joint capsule is a fibrous, elastic envelope that surrounds the joint. The joint capsule and ligaments ensure the joint's stability [3].

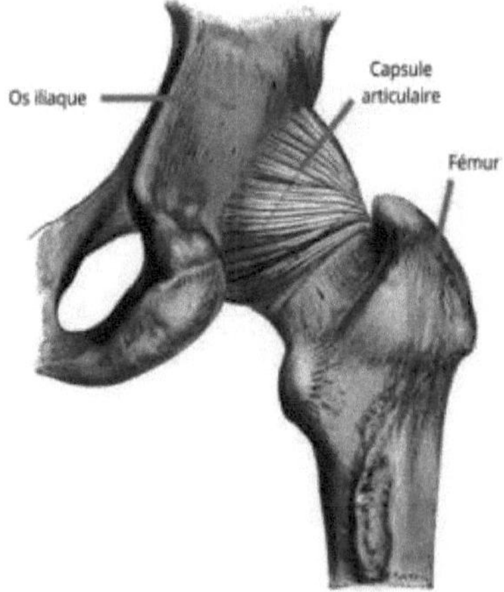

Figure 1- 7: Coxofemoral joint, posterior view [3]

- Femoral head: The femoral head, connected to the upper end of the femur by the femoral neck, represents a portion of a sphere (about 2/3 of a sphere) with a diameter of between 40 and 50 mm, depending on the individual. It is covered by cartilage and the femoral ligament [1].

- The acetabulum: This is a cavity that receives the femoral head (figure 1-8). It is located on the outer surface of the iliac bone. The acetabulum has two parts: the acetabular cavity and the acetabular fossa [1].

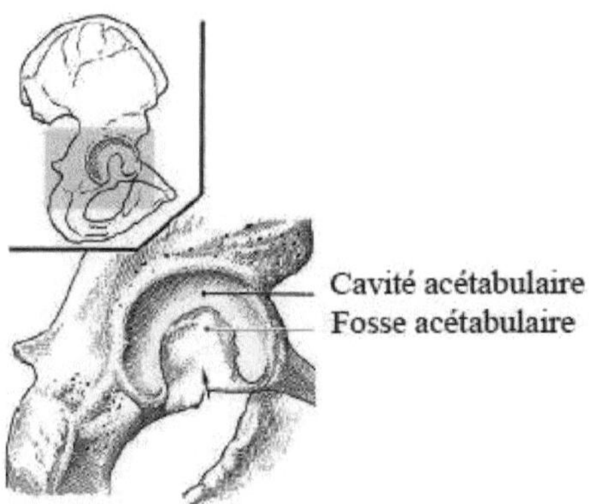

Figure 1- 8: Acetabular cavity (after Kamina) [1]

1.2.2 Joint surfaces

The femoral head articulates in the acetabulum. The contact surface between the head and the acetabulum is slightly larger than the surface of a half-sphere, and is determined by the depth of the acetabulum. This is reinforced by the presence of the acetabular rim (attached to the edge of the acetabulum: figure 1-4) and the transverse ligament. The bony surfaces of the femoral head and acetabulum are covered by a layer of smooth articular cartilage, approximately 3 to 6 mm thick. This cartilage absorbs forces and shocks between the bones, facilitating their sliding against each other (figure 1-9) [2].

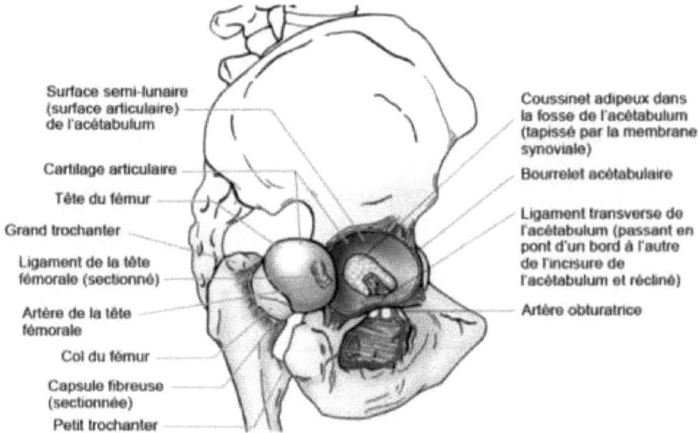

Figure 1- 9: Hip coxofemoral joint, lateral view [2]

1.3 General information on hip replacements

1.3.1 Definitions

A hip prosthesis is an artificial joint replacement for a destroyed hip joint.

Hip prostheses are used in a wide range of conditions, including osteoarthritis of the hip, femoral neck fractures, hip dysplasia and avascular necrosis.

Hip replacement involves surgery to restore joint mobility by creating a new joint space.

Joint surgery is called arthroplasty. In complex arthroplasty, the diseased joint is partially or totally replaced by a prosthesis. When arthroplasty is partial, as in the case of a fractured femur, the head of the femur is replaced by a metal prosthesis of the same shape. In the case of total arthroplasty, both joint surfaces are replaced: in the hip, the head and neck of the femur are replaced by a prosthesis (made of steel, titanium or an alloy of chromium and cobalt), while the corresponding joint surface in the pelvis is replaced by a hollow hemisphere (made of polyethylene or metal) [4].

The aim of this surgery is to relieve pain and improve quality of life for patients undergoing surgery. And it offers patients the chance to return to an active, comfortable life.

1.3.2 History

In the early 20th century[ème] orthopaedic surgeons encountered two types of hip damage: osteoarthritis and femoral neck fractures. Osteoarthritis is the wear and tear of the cartilage, causing the loss of the lining that enables the head of the femur to slide smoothly inside the acetabulum.

To replace the lost cartilage, numerous materials are interposed between the femoral head and the acetabulum: plaster, boxwood, rubber, lead, zinc, copper, gold, silver or fragments of pig bladder, etc. None of these materials is suitable: either too brittle, too soft, too toxic. None of these materials is suitable: either too fragile, too soft, too toxic [5].

The historical development of the total hip prosthesis is summarized in the table below.

Table 1- 1: Chronological evolution of hip replacements [5] and [6].

Year	Author	Description	Figure
1922	Hey-Groves [5]	He replaced the entire femoral head with an ivory sphere of the same caliber, attached by a handle that passes through the femoral shaft. This procedure remains an isolated case, although the result is satisfactory four years after surgery.	*Figure 1-1: Ivory femoral head*
1923	Marius Smith-Petersen [5]	He created the first arthroplasty using thin glass molds that he interposed between the two surfaces of the hip. The lens is just a few millimetres thick. The disadvantage of this method is the fragility and necrosis of the femoral head.	*Figure 1-2: Glass acetabular insert*
1939	Bohlman of Baltimore	He developed the first femoral prosthesis in Vitallium metal (an alloy of 65% cobalt, 30% chromium, 5% molybdenum and	

		other corrosion-resistant substances). It replaces the femoral head and the cartilage that covers it. This solution eliminates the risk of necrosis. He chose to fix the metal head to the outer cortex of the femoral neck with a nail. The first two operations ended in failure, prompting Bohlma to verticalize the nail. The results were somewhat conclusive.	
1946	The Judet brothers [5][6]	They replaced the removed head with a sphere of the same caliber made of methyl methacrylate, better known as Plexiglas. This is attached to a pivot that passes right through the neck of the femur. In all cases, the immediate results are good, but quickly disappointing in the medium term. Changes of shape did nothing to help. These failures were due to an intolerance to acrylic wear debris, which was definitively abandoned in 1949.	*Figure 1-3: Plexiglas head*
1950	Austin Moore [5][6]	He proposed keeping the femoral head supported by a rod fixed in the medullary canal of the femur (figure 1-14). A few days after the operation, the final shape quickly takes on the shape of the stem. The Moore prosthesis is made of Vitallium. A window is made in the prosthetic stem to allow bone regrowth, and a hole is placed at the top of the neck. This solution is very useful for the treatment of femoral neck fractures. In osteoarthritis, however, the worn cartilage of the acetabulum remains unchanged by the metal head. This treatment requires a total prosthesis in which the	*Figure 1-4: Moor prosthesis* Anatomie de l'os *Figure 1-5: Medullary canal*

		femoral head and acetabular cups are replaced.	
1951	George Mac Kee [5][6]	His choice was metal. The new femoral head rolls into the acetabulum, which is covered by a metal shell. Fixation to the bone remains the main problem. The acetabulum is fixed by a large posterior screw. The femoral component is attached to the diaphyseal cortex by a plate. With the stainless steel prosthesis, it can be loosened in less than a year. The Vitallium PTH remained in place for over three years, before the prosthetic neck was broken.	*Figure 1-6: Screw-in cup and head supported by a target plate*
1952	Thompson [5][6]	He proposed a model resembling the Moore prosthesis, but without a window. The Thompson-type model features a femoral component with a slightly smaller head to fit inside the metal prosthetic acetabulum. This model was used from 1956 to 1960. Results were quite satisfactory at over 10 years. However, 10 out of 26 cases suffered loosening.	*Figure 1-7: Thompson prosthesis*
1960	Mac Kee [5][6]	His solution to the problem of repeated friction between one metal part and another was a Vitallium stem carrying a large femoral head articulating in a metal Vitallium acetabulum, and a large acetabular screw holding the two components together. Despite Mac Kee's improvements, early loosening persisted in a large number of cases. Charnley, Mac Kee and Farrar therefore decided to abandon the metal-metal pair and use a high-density polyethylene cup.	*Figure 1-8: Mac Kee total hip prosthesis*

| 1960 | John Charnley [5][6] | He proposed several principles: new materials, cement fixation, new prosthetic head size, new approach.
He has developed a new concept that involves coating the reshaped joint surfaces with a thin plastic film (PTFE polymer).
These thin cups give spectacular immediate results. However, the femoral head that receives the Teflon cup soon suffers dislocation complications. He then went back to his first Teflon acetabular cups and fitted prostheses composed of an acetabulum of his own invention and a Moore-type metal femoral stem, fixed with bone cement.
The results are quite good, but the very thin acetabulum wears quickly and continues to loosen in a significant number of cases. The solution is to reduce the natural diameter of the femoral head from 41 millimeters to 22 millimeters. This is the low-friction arthroplasty. However, after a few years, the small head imposes unbearable pressure on the cup, which wears out far too quickly. | *Figure 1-9: Polymer (PTFE) cup*

Figure 1-10: Low-friction prosthesis |
| 1962 | John Charnley | He chose high-molecular-weight polyethylene. The proposed prosthesis is cemented, with a small 22 mm metal head rolling in a polyethylene acetabulum. But small femoral heads are more easily dislocated. So he proposed a new solution, but even this technique greatly reduces the risk of dislocation. | *Figure 1-11: The Charnley prosthesis* |

1966	Müller [5][6]	Initially, he increased the diameter of the femoral head from 22 mm to 32 mm. The dislocation rate decreased, but wear on the polyethylene acetabulum increased. He therefore changed the diameter to 28 mm. The cemented stem is nicknamed the "banana" prosthesis because of its shape. The acetabulum is made of polyethylene.	 *Figure 1-12: Muller prosthesis*
1970	P. Boutin [5]	He proposes a total hip prosthesis with a ceramic acetabulum and a two-part femoral component: A ceramic head mounted on a steel body.	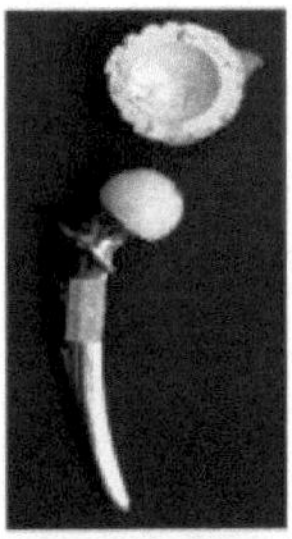 *Figure 1-13: Ceramic femoral head*
1971	Judet [5]	Proposes a direct-anchored prosthesis. He called this cobalt-based alloy "porometal" because the beads covering it were separated by pores. He fitted 1,611 of these prostheses up to 1975, but numerous failures occurred due to the poor mechanical and metallurgical characteristics of the implants.	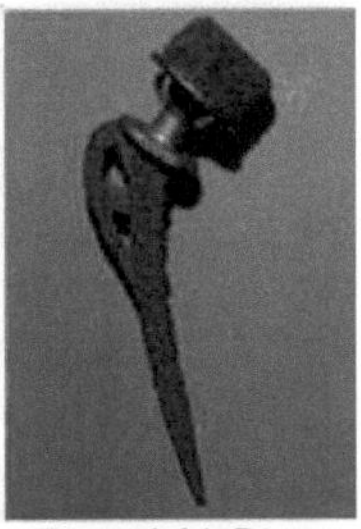 *Figure 1-14: Direct cemented porometal prosthesis*
1977	Engh [5]	He began using a porous metal coating on the femoral stem of prostheses.	 *Figure 1-15: Femoral shaft covered by a porous coating*

| 1979 | Zweimülle r [5] | Presents a femoral prosthesis featuring a pyramidal shape with a rectangular cross-section. | *Figure 1-16: Femoral prosthesis featuring a pyramid shape* |
| 1981 | Bousquet [7] | He developed the concept of dual mobility, and has chosen to use this concept exclusively in all his patients undergoing total hip replacement. | *Figure 1-17: Dual mobility total prosthesis [7]* |

New materials are being researched to provide superior mechanical and tribological properties. Figure 1-27 summarizes the evolution of materials used from the first ivory prosthesis to ceramics [3].

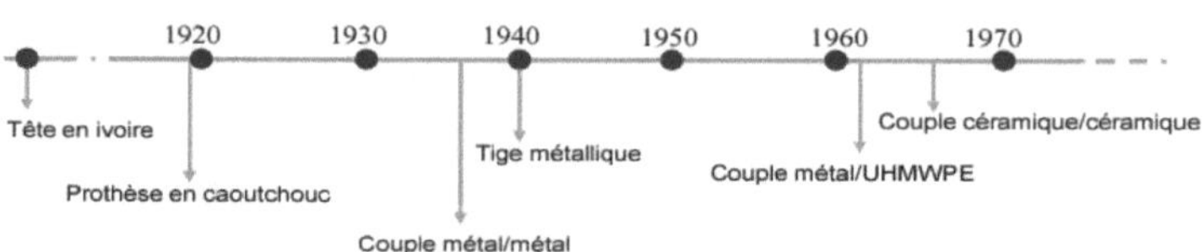

Figure 1-18: History of PTH materials

In 1986, the Corail stem was invented. It's a straight stem entirely coated in titanium and covered with hydroxyapatite [8].

Figure 1-19: Coral femoral stem

In the **2000s**, uncemented stems evolved once again. Their fixation capacity was further improved, and hip misalignment was corrected. These new stems also facilitated direct anterior approach, minimally invasive surgery and implant installation in a narrow femoral shaft with thick cortex. These new stems were a synthesis of Corail's fully-coated straight stem and Muller's self-locking stem designs [7].

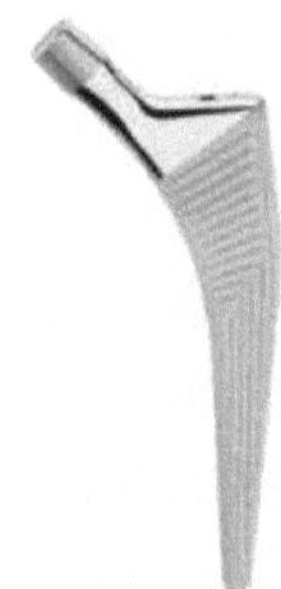

Figure 1-20: Uncemented femoral stem

In 2002, following the abandonment of dual-mobility prostheses due to early errors, a second generation appeared, with results equivalent to the best single-mobility acetabular cups, but above all without dislocation [9].

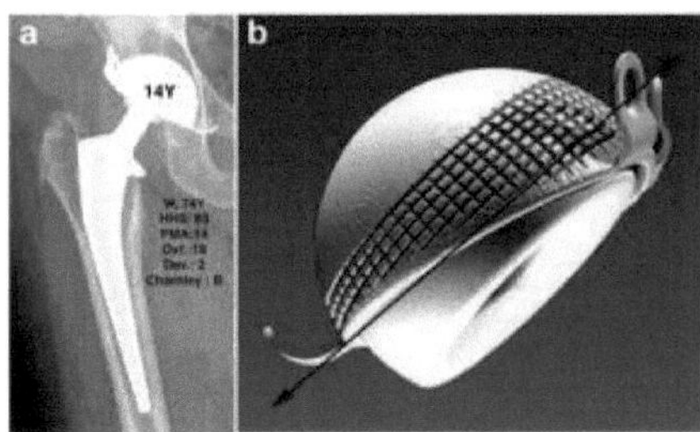

Figure 1-21: Second-generation dual-mobility cup

In 2005, cementless fixation techniques were developed. The femoral stem is covered with a treated surface, enabling it to be integrated into the bone. The solution chosen for the acetabulum is a metal shell impacted into the cancellous bone. As in the case of the femur, its outer surface features miniature reliefs to ensure integration with the pelvic bone [5]. A research group has proposed The method consists in obtaining an approximation of the real 3D volume of the patient's hip by deforming a nominal mesh of the hip in 3D. This method consists of 2 approaches: a study of the exploitable characteristics of the hip and femur for the reconstruction of 3D models, and a methodology for the reconstruction of the 3D volume of the hip using minimally invasive imaging techniques: a single radiographic image (2D) and a few ultrasound images (3D) [10].

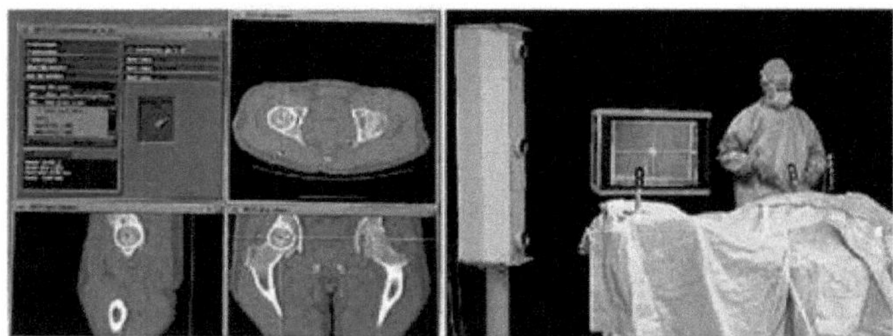

Figure 1-22: 3D ultrasound images

Between 2012 and 2014, 3D radiographic planning evolved as operative planning of a total hip prosthesis on 2D radiographs showed its limitations. This technique made it possible to know the exact dimensions of the joint and plan implants precisely, so surgeons were able to obtain precise control [11][12].

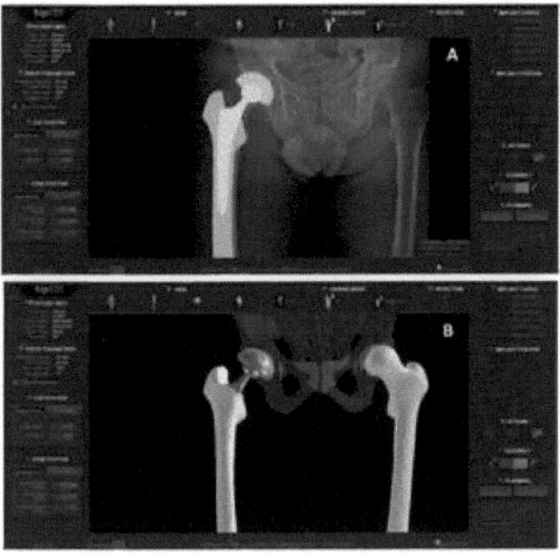

Figure 1-23: Example of 3D computer-aided planning

A Japanese research group has recently developed a new, super-elastic, flexible and resistant metal alloy, which holds great promise for biomedical applications. This cobalt-chromium-based biomaterial mimics the flexibility of human bone and has excellent wear resistance. This new biomaterial could be used for implants such as hip or knee joint replacements and bone plates, reducing the problems associated with conventional implant materials. A new Co-Cr-Al-Si alloy called CCAS. This alloy represents two important innovations for metallic biomaterials: it is the first simultaneous achievement of low Young's modulus and high wear resistance, as well as the first report of enormous superelastic recovery deformation in Co-Cr alloy systems [13].

Newness is not necessarily a good criterion for choosing a total hip prosthesis. A British study [14] found that classic, tried-and-tested models, which are often less costly, are more attractive to people over 65 than newer models.

1.4 Compositions of hip prostheses

The hip prosthesis consists mainly of 3 parts: the cup and insert, the head and the femoral stem (Figure 1-33).

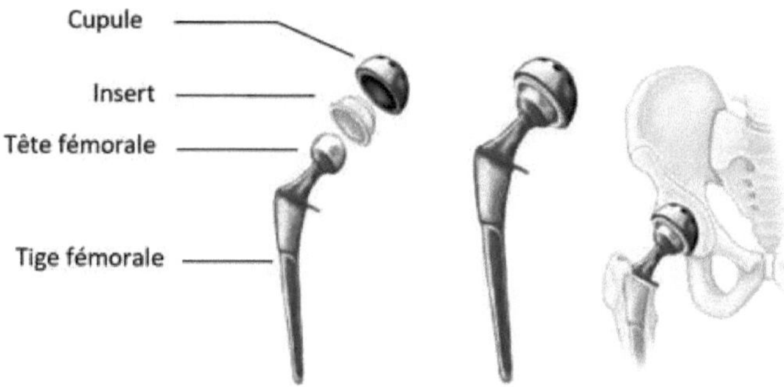

Figure 1-24: Hip prosthesis components

A cup is fixed in the acetabulum of the pelvis. Its central part is shaped like a concave half-sphere and articulates with the femoral head. It can be made of metal, ceramic or polyethylene. The peripheral part is adapted to the type of fixation: metal for cementless prostheses, polyethylene for cemented prostheses (figure 1-34) [4].

Figure 1-25: The cup

It is generally composed of two parts: the metal-back, also known as the acetabular ring: a hemispherical metal part fixed to the pelvic bone; and the insert, usually made of plastic materials and in some cases ceramic (Figure 1-35), fixed in the concave part of the metal-back to fit the femoral head and form the two contact surfaces of the joint together [15].

The insert allows the femoral head to slide more easily and naturally into the acetabulum

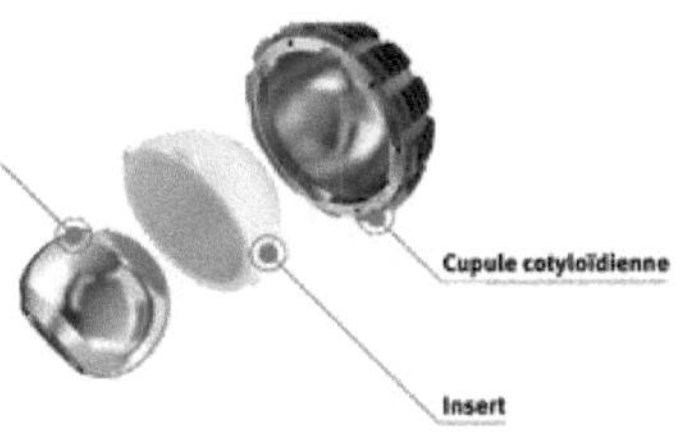

Figure 1-26: Acetabular insert

A femoral head, shaped like 2/3 of a sphere, articulates with the insert to form the prosthetic joint. It can be monobloc with the stem or modular, mounted on a Morse taper at the end of the neck, This head can be made of metal (stainless steel, chrome-cobalt alloys, titanium alloys), ceramic (alumina, zircon) or high-density polyethylene, of variable diameter, and can be changed in the event of exclusive revision of the acetabular component in the absence of femoral loosening (figure 1-36) [16].

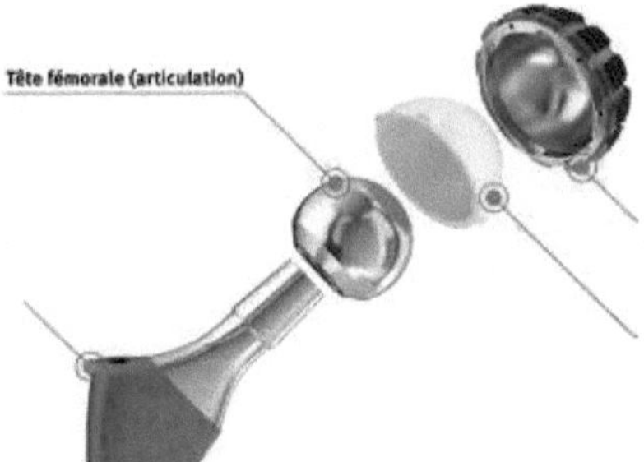

Figure 1-27: Femoral head of the hip prosthesis

A metal **rod** with neck: this is the part that takes up the stresses. It is inserted into the femur after preparation. It can be cemented into the femur with acrylic cement, as in the case of straight stems, or force-fitted into the femoral canal, as in the case of anatomical stems, which follow the morphology of the bone and can be uncemented and coated with hydroxyapatite, or cemented (figure 1-37) [17].

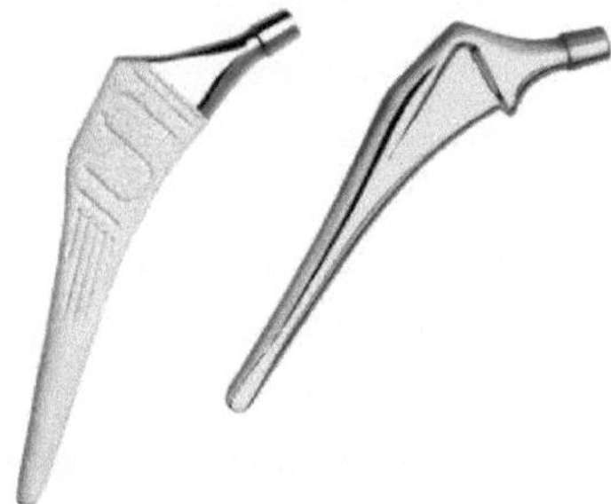

Figure 1-28: Cemented and uncemented femoral stems

1.5 Types of hip prostheses and methods of fixation

1.5.1 Types of hip replacement

Hip prostheses can be either total or partial. The choice of prosthesis is determined by future life expectancy: life expectancy of the implant, early resumption of professional activities, normal relationships.

1.5.1.1 Total hip prostheses

The prosthesis is said to be total when both parts of the hip are replaced: the head of the femur is replaced by a stem ending in a large ball, and the part of the pelvis (the acetabulum) is replaced by a cup. These two parts fit together to form the hip joint. The aim of this prosthesis is to replace the hip joint and enable it to function almost normally [18].

Most often, in a total hip prosthesis, replacement is by an uncemented stem and an acetabulum with an insert that articulates with the head of the stem, which may also be cemented or uncemented.

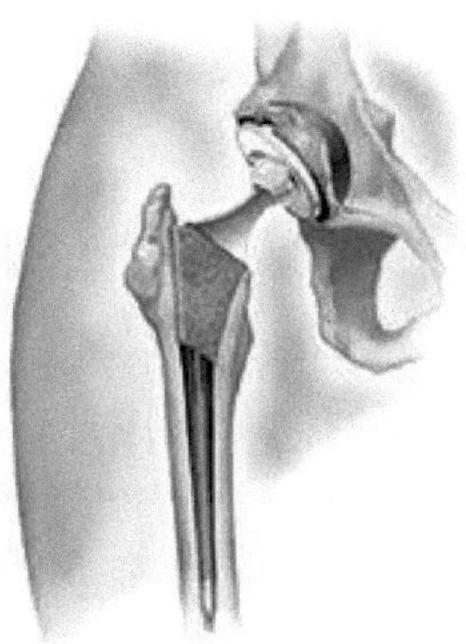

Figure 1-29: Total hip replacement

In the case of total hip prostheses, there is a risk of dislocation, and to avoid this problem, a dual mobility system is used in the prostheses. This system reduces the risk of dislocation, particularly in the posterior track. This dual-mobility system not only restores excellent joint amplitude, but also eliminates the complications associated with total hip prostheses. With conventional prostheses, the dislocation rate ranges from 0.7 to 11%, while dual mobility prostheses have a dislocation rate of 0.01%. This undeniable advantage gives the patient a great deal of freedom, and it is advisable to avoid extreme movements that can lead to pain [19].

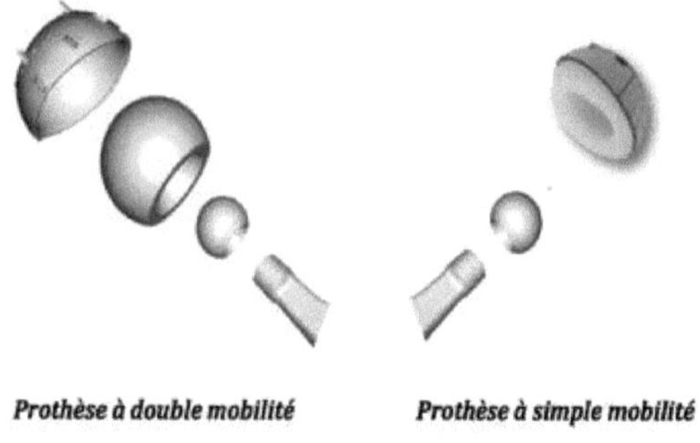

Figure 1-30: The dual mobility prosthesis

1.5.1.2 Partial (cephalic) prostheses

Partial hip replacements replace only the head of the femur. The acetabulum is not replaced. They are most often proposed following fracture of the upper end of the femur [18].

a. One-piece prostheses

The aim of monobloc prostheses is to enable the elderly, who have suffered a fracture of the neck of the femur, to get up very quickly and walk easily, thus avoiding the complications of forced bed rest.

The disadvantage of these prostheses is that they have no orifices on the stem, which causes wear on the acetabular cartilage due to direct friction. This leads to pain and, after a few years, a tendency for the head of the prosthesis to gradually penetrate the pelvis. The use of cephalic prostheses should therefore be reserved for very elderly patients with limited walking ability and limited life expectancy (figure 1-40) [4].

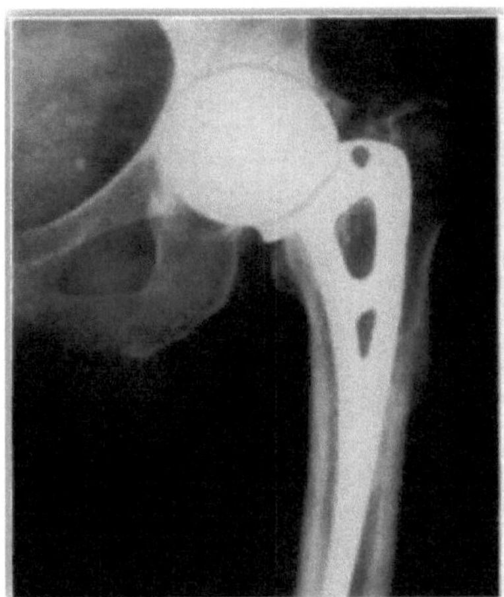

Figure 1-31: Radiograph of a cephalic hip prosthesis

b. Intermediate prostheses

Intermediate prostheses were developed to limit wear on the acetabulum caused by friction with cephalic prostheses. The principle is to create an intra-prosthetic articulation between the head and the tail. During movement, the head of the prosthesis hardly moves in the acetabulum, which limits wear on the cartilage. Mobility occurs essentially in the prosthesis joint.

One of the advantages of intermediate prostheses is that they enable limb length and muscle tension to be adjusted using sleeves or spheres of different depths. The fundamental advantage of this type of prosthesis is that, in the event of acetabular wear, despite prevention by double rotation, the patient can be reoperated on without removing the prosthesis stem, by simply removing the head and sleeve or sphere and fitting a prosthetic acetabulum and a suitable sphere. The intermediate prosthesis is then transformed into a total prosthesis. Cephalic and intermediate prostheses are primarily intended for the treatment of femoral neck fractures in the elderly [4].

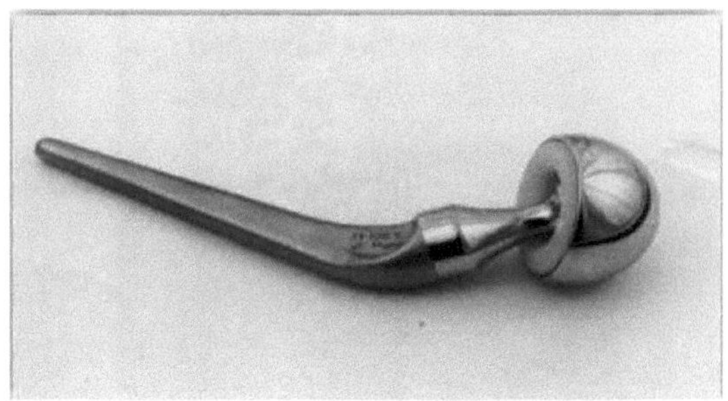

Figure 1-32: An intermediate prosthesis

1.5.2 Fastening systems

Prostheses can be attached to the femur or pelvis using either surgical cement or secondary bone regrowth (cementless).

1.5.2.1 Fixation with cement

This is the most commonly used type of fixation, in which surgical cement is used to fix the prosthetic stem in the femur.

Acrylic-based bone cement is a methyl methacrylate (PMMA) polymer obtained by mixing a powder and a liquid. PMMA is renowned for its excellent optical properties (high refractive index and excellent transparency), as well as for its high chemical inertness, which ensures excellent biocompatibility.

Antibiotic-impregnated cements are used by many surgeons to prevent and treat prosthetic infections. The aim of these medical devices is to allow the antibiotic to spread, thereby preventing bacterial colonization of the implant. Cement has no bacteriostatic effect of its own; it is the antibiotic released by the cement that is solely responsible for antibiotic activity [4].

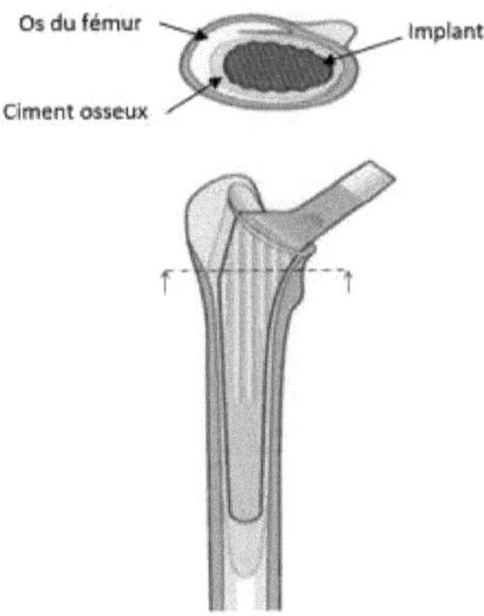

Figure 1-33: The stem structure of a cemented total hip prosthesis [4].

1.5.2.2   Cementless fixation

This is a fixation in which the prosthetic stem is anchored directly in the femur without the use of cement. The prosthesis is covered with a porous coating, enabling the patient's tissues to colonize the prosthesis surface. The coating is an apatitic calcium phosphate whose structure is similar to that of the mineral phase of bone.

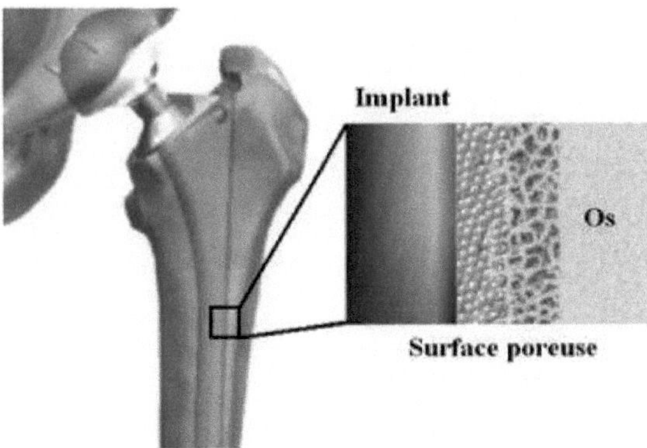

29

1.6 Solicitations

The hip supports multiple loads to maintain balance. And because it is located far from the center of gravity (eccentric), it can support these loads (Pauwels' balance principle) [2].

To determine the stress distribution, we need to know the forces applied to the prosthetic joint during the patient's daily activities. Indeed, several factors can influence stresses, such as patient weight, activities and bone geometry.

1.6.1 Hip movements

The hip joint is assimilated to a ball-and-socket joint with 3 degrees of freedom.

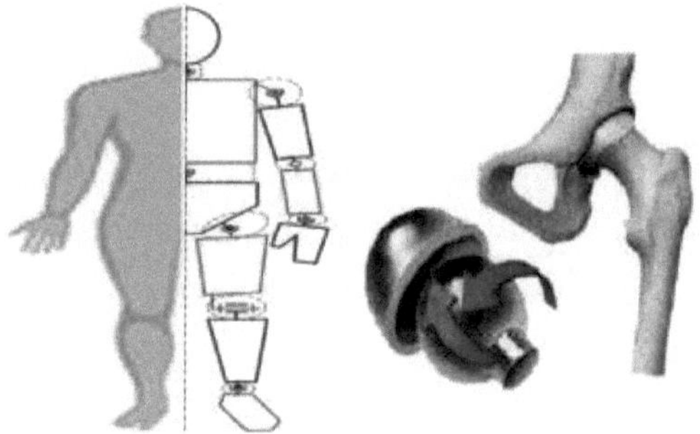

Figure 1-35: Modeling the hip joint

Hip movements are performed by several muscle groups attached to the joint bones by tendons. The most important of these are the gluteal, psoas-iliac and thigh muscles. They operate on three perpendicular axes, giving the hip joint three degrees of freedom of movement, centred on the femoral head [2].

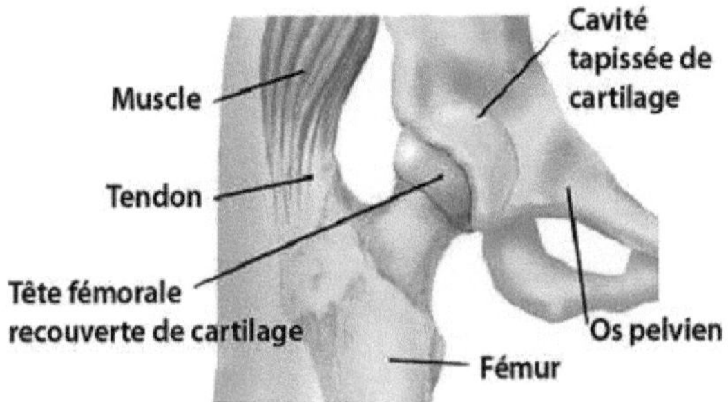

Figure 1-36: Hip muscles

The hip joint allows a variety of simple movements, such as flexion, extension, abduction, adduction, longitudinal rotation and compound movements like circumduction.

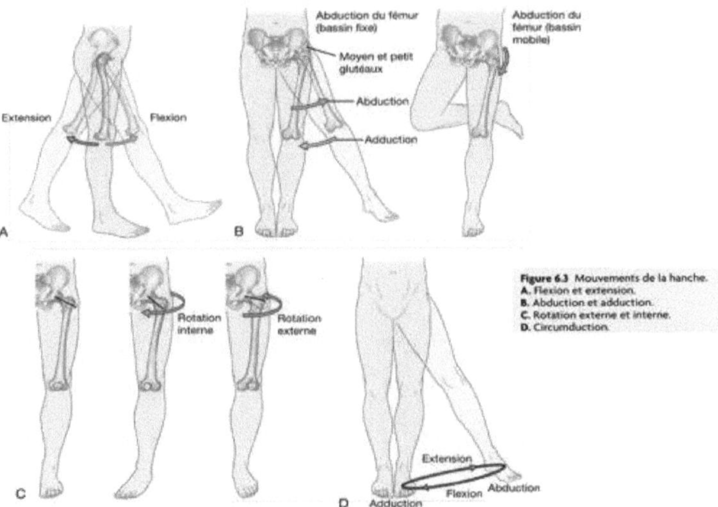

Figure 1-37: Hip movements

1.6.1.1 Flexion and extension

The leg moves forward during flexion, the degree of flexion depending on the position of the knee (up to around 120 and 90 degrees for a bent and straight knee, respectively) (Figure 1-47). During extension, backward movement of the leg is limited (up to 10-20 degrees)[2].

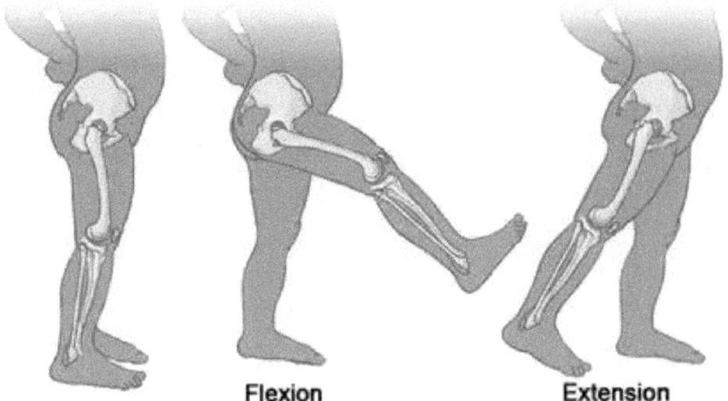

Figure 1-38: Hip movements: flexion and extension (The Hospital for Sick Children, Toronto, Canada, 2004-2012)

1.6.1.2 Abduction and adduction

The leg moves straight out to the side (up to about 35 degrees) during abduction (Figure 1-48). It is returned to its initial position during adduction [2].

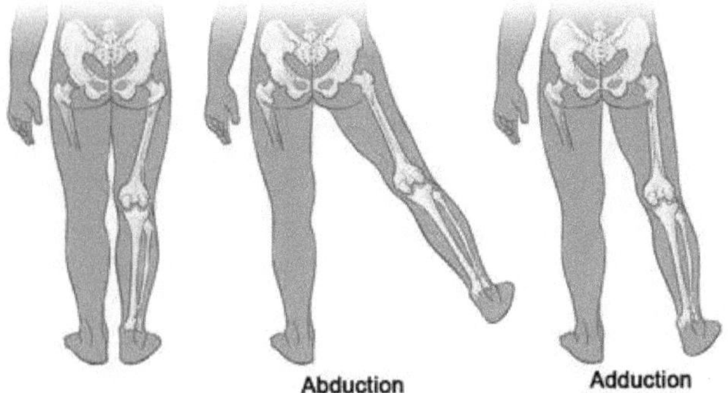

Figure 1-39: Hip movements: abduction and adduction (The Hospital for Sick Children, Toronto, Canada, 2004-2012)

1.6.1.3 Medial and lateral rotation

The thigh is turned inwards during medial rotation (up to 35-40 degrees) and outwards during lateral rotation (up to 45-60 degrees) [2] (Figure 1-49).

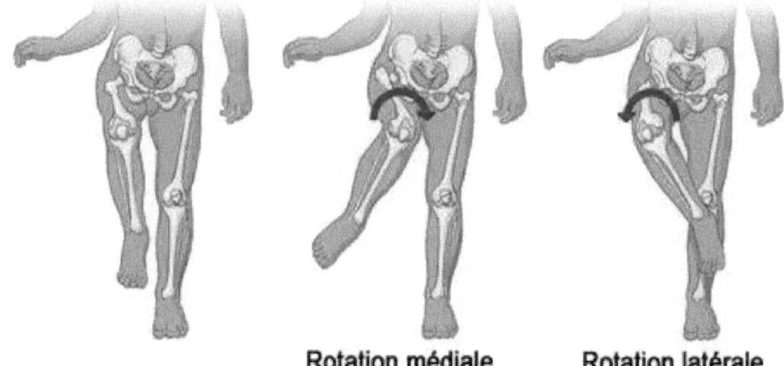

Figure 1-40: Hip movements: medial and lateral rotations (The Hospital for Sick Children, Toronto, Canada,2004-2012)

1.6.1.4   Longitudinal rotation

Longitudinal rotation of the hip takes place around the mechanical axis of the lower limb. In the straight position, this axis merges with the vertical axis of the coxofemoral joint. Under these conditions, external rotation carries the ball of the foot outwards, and internal rotation carries the ball of the foot inwards [1].

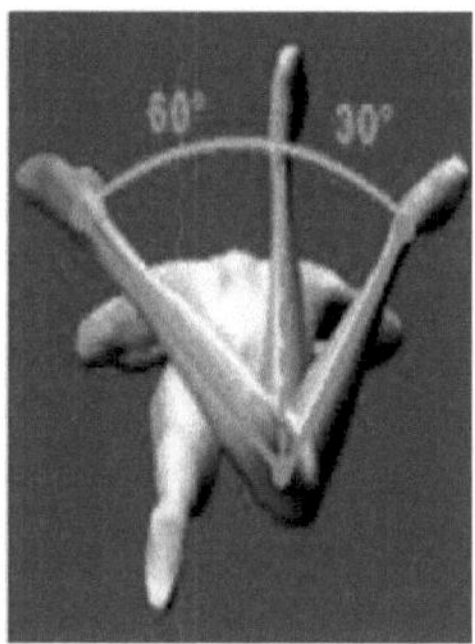

Figure 1-41: Longitudinal rotation movement

1.6.1.5   Circumduction

As with all joints with three degrees of freedom, hip circumduction is defined as the combination of elementary movements simultaneously around the three axes. When circumduction is pushed to its extreme amplitude, the axis of the lower limb describes a cone in space, the apex of which is occupied by the center of the coxo-femoral joint: this is the circumduction cone [1].

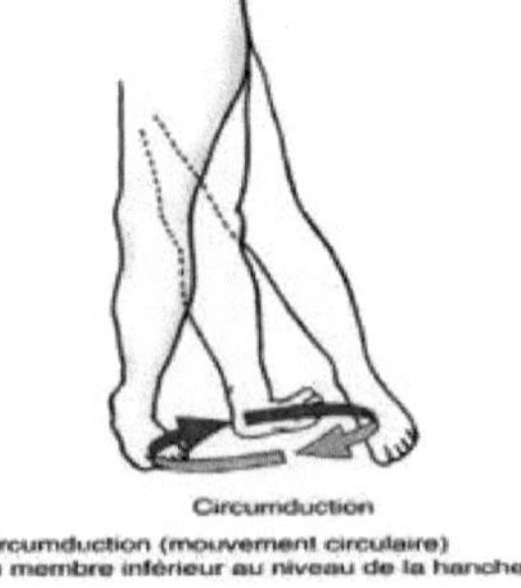

Figure 1-42: Circumduction movement

1.6.2 Hip mobility

The stability of the human skeleton largely depends on the persistence and functional quality of the hip joint. From a biomechanical point of view, there are three degrees of freedom activated around three axes specific to the hip joint.

Anteroposterior axis, ensuring abduction and adduction (AB/AD).

Vertical axis merged with the longitudinal axis of the lower limb, enabling external rotation and internal rotation movements (RI/RE).

Transverse axis, around which flexion and extension (F/E) take place [20].

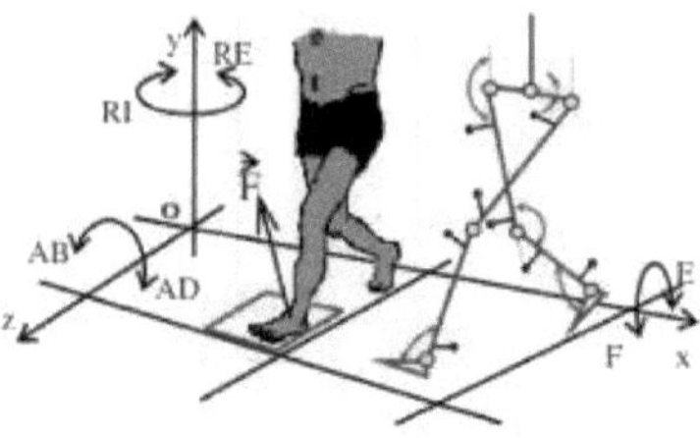

Figure 1-43: Resultant transmitted to the hip in monopodal support position [20]

1.6.3   Loads applied to the hip

Determining the internal forces present during human movement is essential to understanding the body's mechanical function, in order to assess the loads and risks to which the human body is subjected. Hip stresses are such that, in a bipodal situation, half the body weight is distributed on each hip. Whereas in a unipodal position, the concept of the "Pauwels balance" shows that the hip joint supports around 3 times the body weight (figure 1-53). Under normal conditions, the cartilage cushions and distributes the transmission of forces within the joint [3].

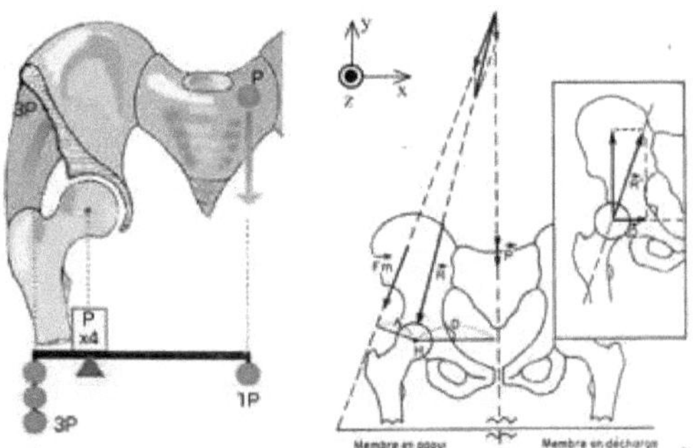

Figure 1-44: Schematic of the Pauwels balance (Hertogh, 2013) [20],[2]

For equilibrium to exist, the moments of the concentrated forces must be equal, i.e. $PxD = F_m xh$. Since anatomical measurements are defined by $D = 3h$, $F_m = 3P$. In total, the resultant R is equal to the algebraic sum of the two forces, i.e. $R = 4P$ [20].

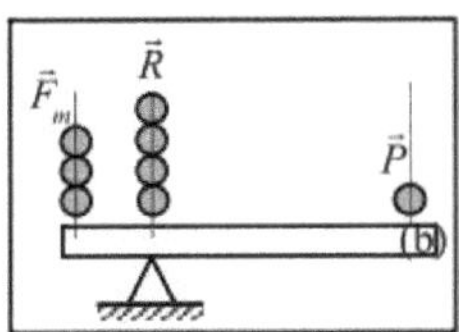

Figure 1-45: Force diagram

Certain anatomical variations in the femur (coxa valga, short neck) considerably increase this force and thus the pressures borne by the hip. Moreover, the smaller the contact surface between acetabulum and femur (insufficient acetabular coverage), the greater the coxofemoral pressures.

1.7 Fatigue and cracking of implantable rods

Determining the mechanical properties of a material, whether static or fatigue, requires laboratory testing. These tests are based on the loads, environment and shape of the structure.

1.7.1 Fatigue test

The aim of fatigue testing is to determine the ultimate strength for a given load and duration. Fatigue tests are carried out on specimens subjected to loads of constant amplitude until crack initiation is observed.

The number of cycles to initiation obtained experimentally is then compared with the number of cycles to failure of the structure. These tests are repeated for different levels of loading amplitude to establish the material's Wöhler curve, giving the stress amplitude noted σa or σ as a function of the cycle at initiation or failure (in logarithmic scale).

These tests are generally carried out in accordance with international technical standards voluntarily applied by manufacturers.

The following table shows the various fatigue tests on hip prostheses according to the ISO standard:

Table 1- 2: Fatigue tests to ISO standard

Testing	Reference standards

Dimensional measurements	ISO 7206-2
Rod fatigue tests (Low stem cementing)	ISO 7206-4
Fatigue tests on head/neck assemblies (High stem cementing)	ISO 7206-6

ISO 7206 describes the apparatus and test procedure for assessing the fatigue strength of femoral stems, mainly in two different configurations, the first with the stem cemented to a predefined distance from the center of the head, the second with the stem cemented to the osteotomy plane.

1.7.2   Fatigue tests on the ISO 7206-4 femoral stem

Example of a fatigue test on a rod (low cementing of the rod): Fatigue tests carried out by Yunus E Delikanli1 and Mehmet C Kayacan on Ti-6Al-4V titanium superalloy rods [21].

The tests are carried out on a servo-hydraulic fatigue testing machine (Instron 8872) with a load capacity of 25 kN and a torque of 100 N m at room temperature. Acrylic-based bone cement, frequently used in total hip prosthesis and resistant to dynamic testing, is used to fix the implant in the defined position. The specimens are positioned in accordance with the standard mentioned, providing maximum load in the proximal region of the femur and allowing the stem to be loaded.

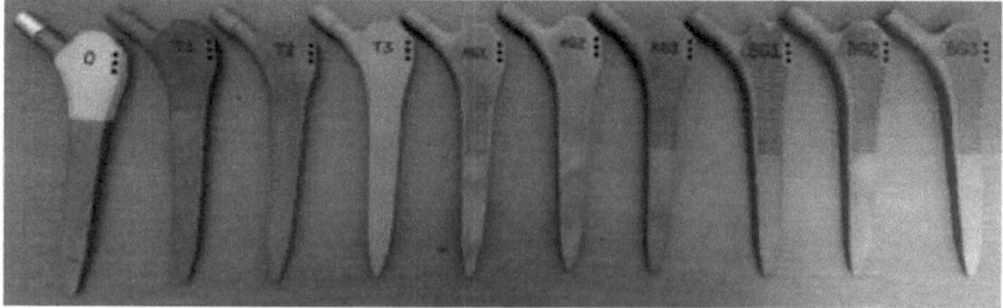

Figure 1-46: Specimens prepared for fatigue testing

The properties of the materials used in the fatigue test are shown in the following table:

Material	Fish coefficient	Modulus of elasticity (GPa)
Steel (suppression blocks)	0.3	210
Ti6AI4V (femoral stem)	0.342	110
Cr-Co (femoral head)	0.33	200
Bone cement	0.3	3.8

A fixture (clamp) illustrated in Figure 1-56 is used for positioning at the appropriate angles. The specimens are then placed on the clamp in a stainless steel cup into which bone cement has been poured.

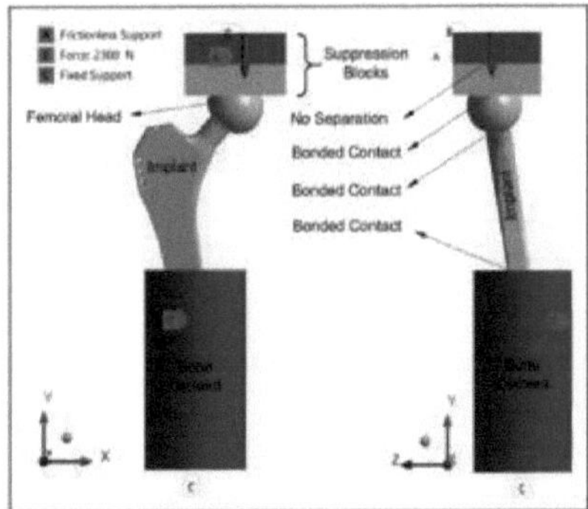

Figure 1-47: Fatigue test finite element analysis model

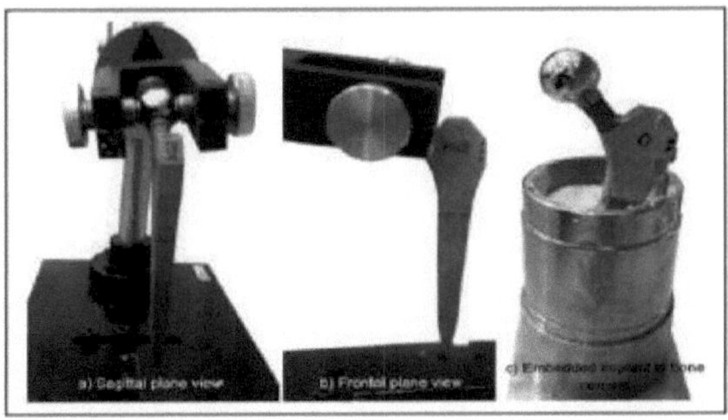

Figure 1-48: Prosthesis in the fixation clamp (a) and (b); stem embedded in bone cement (c).

During the fatigue test, the prosthesis is subjected to sinusoidal compressive loads (with a frequency of 15 Hz and limits between 300 N and 2300 N). Vertical displacements of the specimens are measured by a displacement transducer connected to an actuator in the test machine every 50,000 load cycles. Detailed test conditions are given in the following table:

Table 1- 4: Fatigue test conditions

Frequency	Max. load	Min. load	Temperature	Rod embedding
15 Hz	2300N	300N	Ambient (22°)	With bone cement

The results of the vertical displacement and maximum equivalent stress, from the finite element analysis for each, are shown in Table 1.5 :

Table 1- 5: Finite element analysis results

Specimen pore size (mm)	0.3	0.5	0.6	0.7	0.8	1
Displacement (mm)	0.24214	0.24946	0.25308	0.25910	0.26548	0.28187
Equivalent stress (MPa)	232.63	234.73	236.39	237.85	238.460	240.040

1.7.3   Damage mechanics

Damage mechanics is based on the idea that discrete defects appearing in the material under load (cracks, porosity openings) can be continuously integrated into the macroscopic behavior of the damaged material. A scalar field variable, the damage D, is used to describe the state of the material: $D = 0$ implies that the material is intact and $D = 1$ that it is ruined [22].

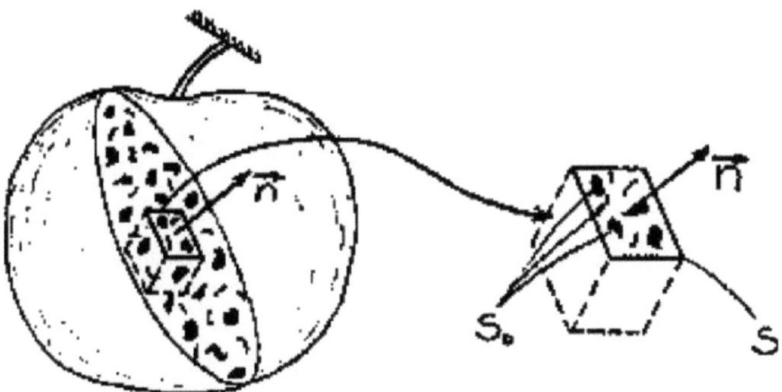

Figure 1-49: Lemaître and Chaboche's damage variable [23]

1.7.4 Cracking

Fatigue cracking occurs when the chosen damage variable "D" reaches its critical value (often 1). The most widely used today is the Palmgren-miner linear damage accumulation law, which remains the best compromise between simplicity of application and quality of prediction for long service lives [24].

The Palmgren-Miner rule is a linear cumulative damage assumption and is the best estimate of the fatigue limit available to date. It states that the sum of the ratio between the number of cycles at a given amplitude and the number of cycles to failure is equal to the accumulated damage.

$$Dp = \sum_{i=1}^{k} \frac{ni}{Ni}$$

Where ni is the number of cycles under a constant amplitude, Ni is the total number of cycles to failure at this amplitude, and Dp is the damage parameter.

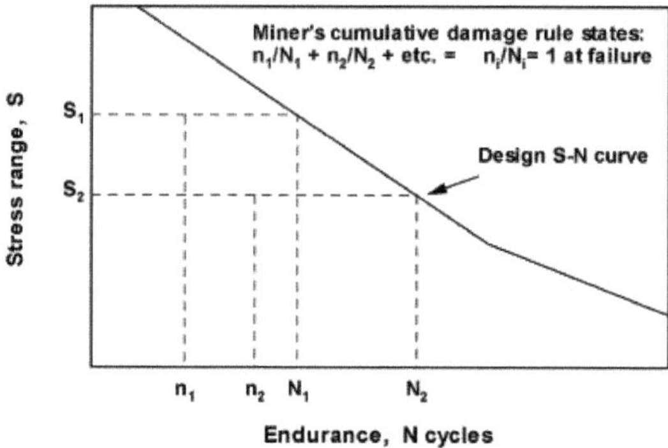

Figure 1-50: Palmgren-Miner diagram

In the field of fatigue modeling, Miner's linear cumulative law is still the most widely used, despite its many drawbacks. This is because it allows us to predict the failure of a part under variable loading by calculating the damage to the specimen.

1.8 Manufacturing materials for hip prostheses

The choice of materials used affects function, which is why biomaterials are used to replace a part or function of an organ or tissue, because biomaterials adapt better to the human body.

Researchers have developed multiple natural and synthetic biomaterials, some of which are used clinically: biopolymers, ceramics and composites. The advantage of these synthetic biomaterials is that they can be synthesized in large quantities, with controlled chemical and physical properties.

For hip prostheses, several materials are used to adapt to the human body.

1.8.1 Classic materials

The materials used for prostheses are characterized by 3 essential criteria: biocompatibility (good tolerance by the human organism), corrosion resistance and mechanical properties.

Polyethylene parts: Acetabular inserts

Ceramic parts: prosthesis femoral heads and inserts

Metal parts: Femoral stems, heads and inserts

1.8.1.1   Polyethylene

It remains the most widely used material, thanks to its low cost and ease of fabrication. Indeed, cemented fixation of polyethylene directly into the bone ensures better wear resistance. However, even when ideally cemented and of sufficient thickness, polyethylene is subject to wear phenomena. Cross-linking is a tricky process, requiring optimum control of the manufacturing process. If the wear resistance of this type of polyethylene is confirmed over time, it will reinforce the use of cemented polyethylene [25].

1.8.1.2   Ceramics

The origin of the term "ceramic" comes from the Greek meaning clay or potter's clay. It is a fired clay which, after exposure to heat, undergoes a definitive transformation. In contemporary hip prostheses, friction torque has become the main factor in prosthetic longevity [3].

Ceramics are non-organic, non-metallic solids. There are several types of ceramic, but alumina and zirconia are generally used in orthopedics. The manufacture of ceramics requires complex technology. Bioactive ceramics such as hydroxyapatite (a natural component of bone) are also used. Other materials of biological origin, such as coral, a natural porous ceramic, are rarely used.

- Alumina: Alumina is one of the most widely used ceramics, thanks to its special properties: very high hardness, high tensile strength, chemical inertness and density of the order of 4. It is a stable single-phase material with a single type of molecular arrangement.
- Zirconia: mechanical properties four times those of alumina). Zirconia (ZrO_2) crystallizes in three types of lattice: monoclinic (M), cubic (C) and tetragonal (T), which change with temperature, manufacturing processes and additives.

This type of material offers very interesting mechanical properties for clinical use: excellent biocompatibility, very high hardness and excellent lubrication properties.

1.8.1.3 Metal

The metals used in hip prostheses are in fact alloys combining two or more components, at least one of which is metallic. They include stainless steels, chromium-cobalt alloys, titanium-based alloys and the new nickel-titanium alloys;

1.8.1.4 Friction torques

Since the first implantations of total hip prostheses, numerous technical possibilities have been developed to improve the friction torque between the head and the acetabulum of a total hip prosthesis.

The materials used must withstand the stresses of bearing, and have sliding properties (hardness, wettability, roughness, etc.) that do not cause wear debris in the joint. In fact, it has been shown that rubbing metal heads against a polyethylene acetabular cup leads to wear debris, which can cause loosening or osteolysis (bone destruction).

Researchers therefore modified or replaced the acetabular polyethylene and metal head with other materials to reduce wear and thus the risk of revision surgery.

There are many materials that can be used to ensure sliding between acetabular and femoral implants, and numerous combinations are possible. The main friction torques used are :

- Polyethylene-Metal ;
- Polyethylene-Ceramic ;
- Ceramics-Ceramics ;
- Metal-Metal [2].

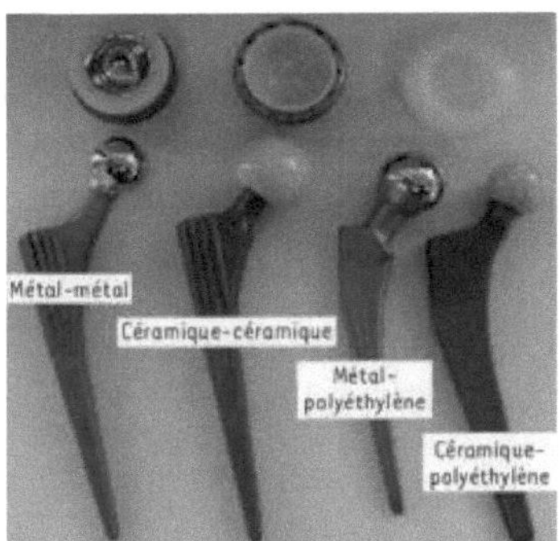

Figure 1-51: Friction torque of total hip prosthesis

The following table summarizes the advantages and disadvantages of the different friction torques.

Table 1- 6: Advantages and disadvantages of friction torques [25][26].

Friction torque	Benefits	Disadvantages	Figures
Polyethylene - Metal	-Life expectancy Polyethylene is highly impact-tolerant -Suitable for osteoarthritis in young people, particularly those suffering from hip dislocation, and for people over 65, even those who are very active, -Reliability and longevity that will make this prosthesis definitive [25].	-Polyethylene deforms and wears with friction -Faster wear and loosening of the acetabulum -Smaller heads dislocate more easily [25][26].	*Figure 1-52: Polyethylene-metal coupling*
Polyethylene - Ceramic	Ceramic has a smooth, very hard, abrasion-resistant sliding surface, reducing wear on polyethylene. -It is inert and chemically very stable, with excellent corrosion resistance [25].	-Ceramic is a brittle material. -To avoid this risk of fracture, the femoral stem cone and the head must come from the same manufacturer to be perfectly matched. -A high price	*Figure 1-53: Polyethylene-ceramic couple*

		-High risk of edging and loosening when polyethylene is cemented into bone	
Ceramics - Ceramics	-Very low wear resistance Bio-inert nature of wear debris -Low friction [26]	Risk of implant fracture -Requires metal-back insertion for excellent fixation. -Cher. -Its use is recommended for the most active subjects. [25]	*Figure 1-54: Ceramic-ceramic coupling*
Metal-metal	-Excellent wear resistance, superior to that of a polyethylene ceramic couple in vivo. -Potentially longer service life than polyethylene due to reduced wear. -Lower dislocation rate. -Use in the most active subjects.	-Possible carcinogen. - Effect of metal ions.	*Figure 1-55: Metal-metal coupling*

1.8.2 Titanium

Titanium is a metallic material with excellent corrosion resistance and high specific mechanical strength (ratio of tensile strength to density). Its density is equivalent to around 60% that of steel. It also has a low modulus of elasticity of 110 GPa (half that of stainless steel at 220 GPa). These characteristics make it particularly interesting for many applications. In addition to these characteristics, its chemical biocompatibility makes this material a prime candidate for the medical field [27].

1.8.2.1 Titanium alloys

Titanium alloys have given medical research the possibility of using a material with a Young's modulus close to that of bone, with other properties such as high corrosion resistance due to the formation of an oxide layer on its surface, optimal weight, wear resistance and a good weight/density ratio, thus correcting defects present in other materials and reducing the risk of failure [28]. They offer biological and mechanical biocompatibility, which is why they are used for prostheses. Currently, titanium-based alloys, in particular Ti-6Al-4V and Ti-6Al-7Nb, are the most commonly used materials for joint prostheses, being registered in the ASTM standard as biomaterials [29].

Today, additive manufacturing is seen as a new impetus for innovation and growth in medical prosthetics, making it possible to produce tailor-made implants.

Titanium alloys are classified into three categories: α alloys, α/β alloys and β alloys.

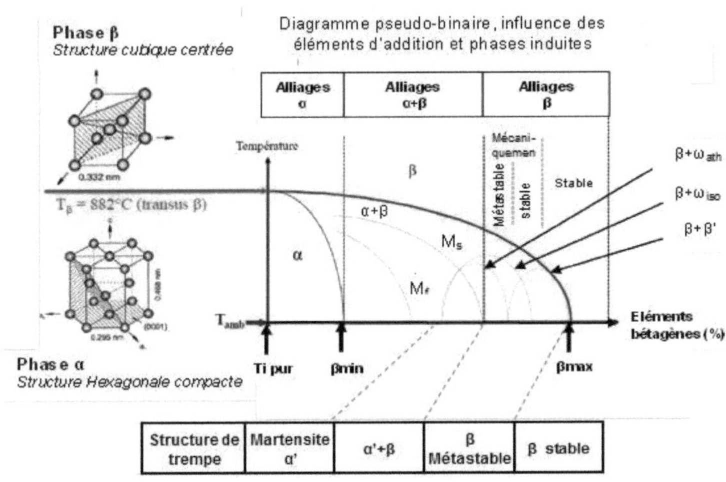

Figure 1-56: Pseudo-binary diagram [27]

α-type alloys: Consisting of 100% α phase. These alloys do not permit structural hardening. They are not very sensitive to heat treatment, their forming is difficult. Their main advantages are low density, good creep resistance and good weldability.

$\alpha+\beta$ type alloys: Consisting of highly variable proportions of α and β phase. They represent the vast majority of titanium alloys available on the market. The best-known is the Ti-6Al-4V alloy (commonly known as TA6V) used in the manufacture of total hip prostheses. Their main advantages are their high mechanical properties, ductility and resistance to oxidation.

β-alloys: Consisting of 100% β-phase. A distinction can be made between β-metastable and β-stable alloys, they have excellent ductility and average strength in the hardened state [27][8].

1.8.2.2 Mechanical properties of alloys

Table 1- 7: Mechanical properties of alloys [27][30].

	Modulus of elasticity E (GPa)	Yield strength σe (MPa)	Tensile strength σmax (MPa)	Fatigue strength σfin, (MPa)
Titanium (TI-CP)	110	170-485	20-550	300
Ti-6Al-4V	113-118	795-880	860-1000	620-800

Material	Standard	Modulus (GPa)	Tensile strength (Mpa)	Alloy type
First generation biomaterials (1950–1990)				
Commercially pure Ti (Cp grade 1–4)	ASTM 1341	100	240-550	α
Ti–6Al–4V ELI wrought	ASTM F136	110	860–965	α+β
Ti–6Al–4V ELI Standard grade	ASTM F1472	112	895–930	α+β
Ti–6Al–7Nb Wrought	ASTM F1295	110	900–1050	α+β
Ti–5Al–2.5Fe	–	110	1020	α+β
Second generation biomaterials (1990–till date)				
Ti–13Nb–13Zr Wrought	ASTM F1713	79–84	973–1037	Metastabe β
Ti–12Mo–6Zr–2Fe (TMZF)	ASTM F1813	74–85	1060–1100	β
Ti–35Nb–7Zr–5Ta (TNZT)		55	596	β
Ti–29Nb–13Ta–4.6Zr	–	65	911	β
Ti–35Nb–5Ta–7Zr–0.40 (TNZTO)		66	1010	β
Ti–15Mo–5Zr–3Al		82		β
Ti–Mo	ASTM F2066			β

Figure 1-57: The evolution of titanium alloys for medical applications

1.9 Manufacturing methods

Various manufacturing techniques have been put into practice to produce hip
femoral stems

1.9.1 Machining: 5-axis machine

The criteria required by the medical industry include machining complex
geometric shapes, with impeccable surface finishes and high dimensional accuracy.

More often than not, these parts require noble materials such as titanium,
chrome-cobalt, stainless steel, ceramics or plastics.

The orthopaedic parts manufactured by the 5-axis machine are : Prosthetic
joints, Modular collars, Cupules, Tibial bases, Cotyles, Mobile inserts, Femoral
stems

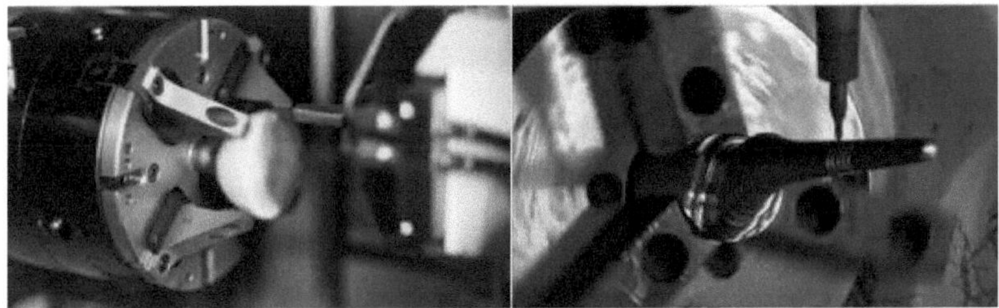

Figure 1-58: Machining on a 5-axis machine

1.9.2 Forging

Forging is recognized as the best technique for manufacturing prostheses, thanks
to its exceptional properties in terms of mechanical strength.

The hip prosthesis is the most mechanically stressed implant in the human body.
The neck of the hip stem (corresponding to the femoral head before the implant is
fitted) is subjected to extremely high fatigue loads, up to several tonnes per cm^2
during certain walking or jumping movements.

The forge fibers the internal metal structure of the orthopedic stem. This
fibering, which follows the curves of the implant, ensures greater fatigue resistance
than any other production technique currently on the market.

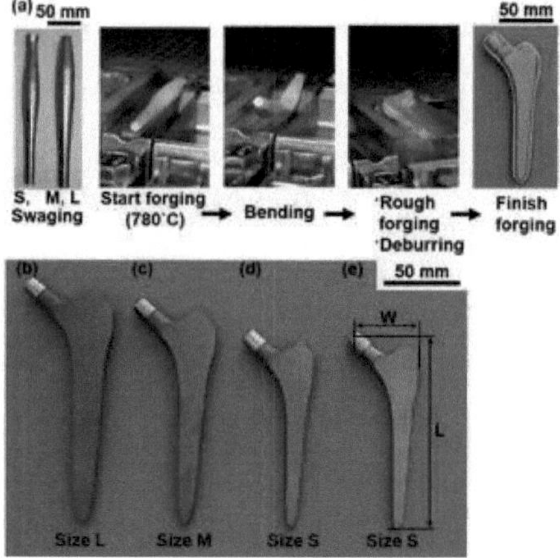

Figure 1-59: Forging process

1.9.3   3D printing using additive manufacturing (AM)

Advances in additive manufacturing (AM) allow us to materialize ideas and designs previously thought impossible to manufacture. Combined with the spirit of biomimetics, which can be described as the art and science of imitating biological systems or nature in general.

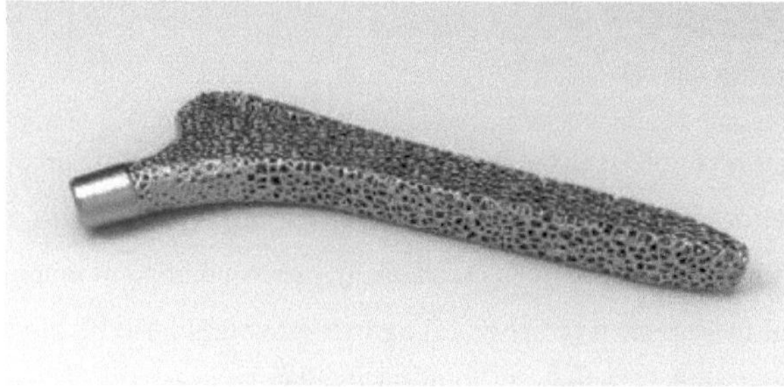

Figure 1-60: Design of an FA-optimized Ti-6Al-4V hip prosthesis stem [31].

Additive manufacturing (AM) enables the designer to create three-dimensional models that can feature diffused porosity in the area of contact with bone, which could ensure a far superior structural and functional connection between living bone and the surface of an artificial implant (osseointegration).

Medical device manufacturers can design implants that mimic the stiffness and bone density of the patient. 3D-printed implants can reduce bone density protection and further improve physical function [32].

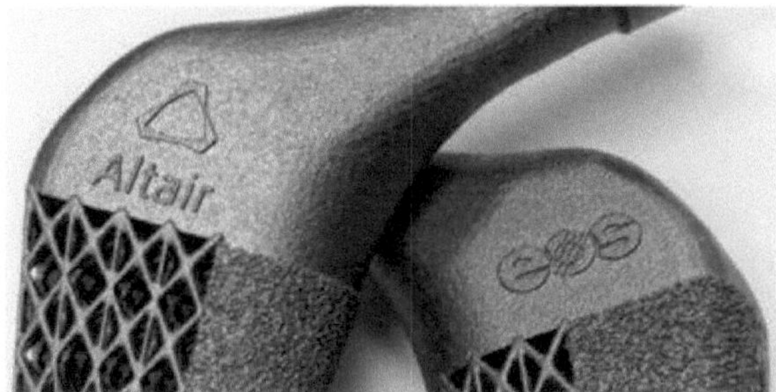

Figure 1-61: 3D printed implant made of Ti-6Al-4V

In the optimized design, stresses in the stem were kept below 575 MPa for all load cases, including jogging. For the titanium sample for which the implant was designed, this equates to an endurance limit of around 10,000,000 cycles.

Figure 1-62: Solid mesh hip prosthesis printed with Ti-6Al-4V

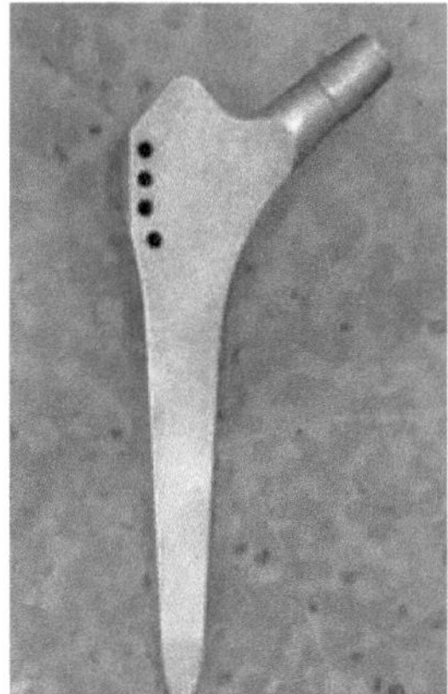

Figure 1-63: Femoral stem manufactured in CETIME

The prosthesis has to match the patient's dimensions as closely as possible, and these vary from person to person due to age, gender, bone structure and morphology. The standard prosthesis cannot match the exact dimensions of each patient, causing post-operative problems such as pain or unevenness in limb length. The 3D printer, by creating customized models, can avoid these problems.

For certain pathologies, it is not always possible to fit a prosthesis.

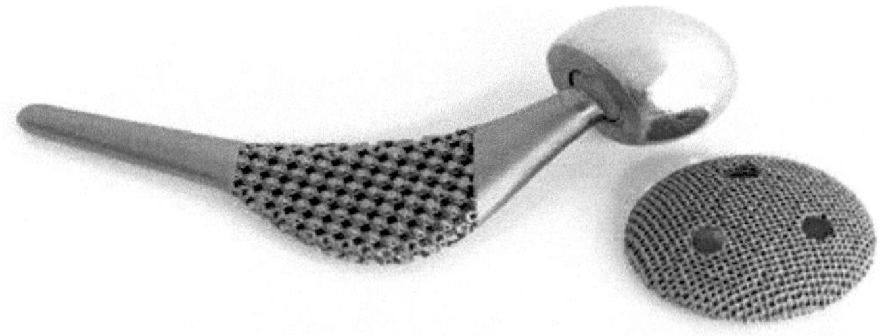

Figure 1-64: Total hip prosthesis made by a 3D printer

Electron beam fusion (EBM) 3D printing manufacturing of patient-specific biomedical implants, replacing knee and hip systems, using additive manufacturing (AM) of Ti-6Al-4V powders by electron beam fusion (EBF) [33].

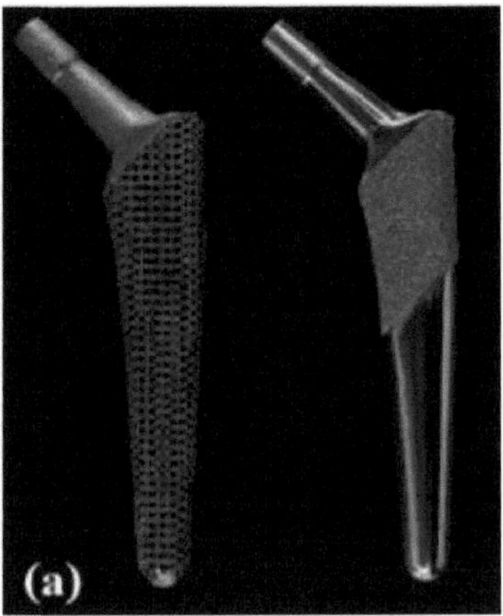

Figure 1-74: Electron beam melting (EBM) printed femoral stem

1.10 Conclusion

This chapter clarifies the concepts and generalities of a hip prosthesis, helps you understand how it works, follow its evolution over time and study the materials used.

Surgical intervention is a practical and effective solution, especially for young, active patients. However, damage to the prosthetic implant can lead to complications such as the need for a second revision operation. To avoid the risk of fracture, digital tests are carried out using the MEF method.

CHAPITRE 2: STATE OF THE ART ON STRENGTH SIMULATION OF HIP IMPLANTS

2.1 Introduction

In this chapter, we will present a review of the literature concerning the most relevant work that has been carried out to study the strength of hip prosthesis femoral stems.

2.2 Studies on titanium hip implants using finite element simulation

This study, presented by Abhinav Kumar, Apoorv Rathi, Jagjit Singh and N. K. Sharma [28], focuses on identifying the most suitable material for hip joint implants with minimum stress, strain and weight, using CATIA V5 software for design and ANSYS Workbench 14.5 for finite element simulation. Six titanium alloy materials were used: Ti-6Al-4V, Ti-5Al-2.5Fe, Ti-6Al-7Nb, Ti-12Mo-6Zr-2Fe, Ti-15Mo and Ti-13Nb-13Zr. All these entities for each material were obtained and compared in order to propose the most suitable material for the manufacture of a stem implant.

2.2.1 Modeling

The hip joint implant was modeled using CATIA V5 software. This is the most commonly used size. The model of the hip prosthesis is shown in figure 2-1.

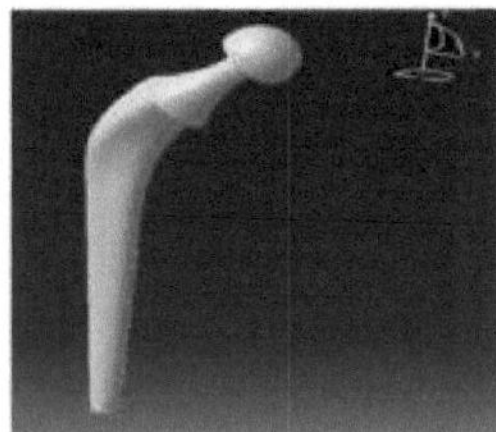

Figure 2- 1: 3D model of a hip prosthesis

2.2.2 Simulation

The hip joint implant has been modeled as a solid piece of metal. The main function of the implant stem is to support the body's load when inserted into the femur, and it is designed to behave like a whole bone. To achieve results similar to those obtained under real-life conditions, the implant's boundary conditions are kept similar to those of a whole femoral bone. Stress distribution and deformation are studied for a 75 kg man in normal standing position. A pressure of 750 Pa is applied to the femoral head of the stem, and the stem base is fixed. The side walls of the femoral stem are provided with a friction-free support to obtain conditions similar to those in a real environment.

2.2.3 Results and interpretation

The finite element model of the hip joint implant was subjected to a pressure of 750 Pa and investigated for equivalent stresses, strains and weights.

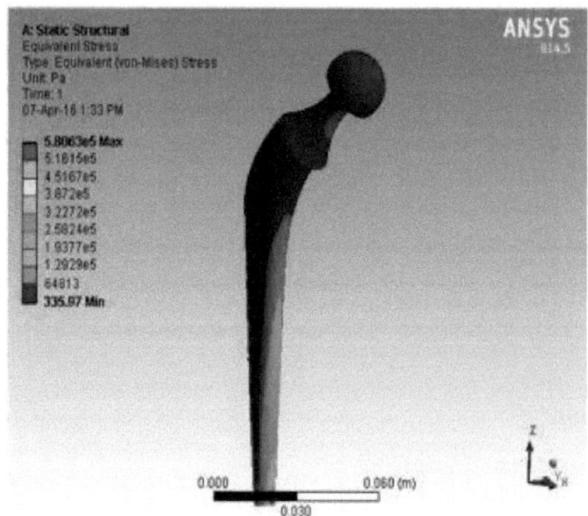

Figure 2- 2: Equivalent constraints

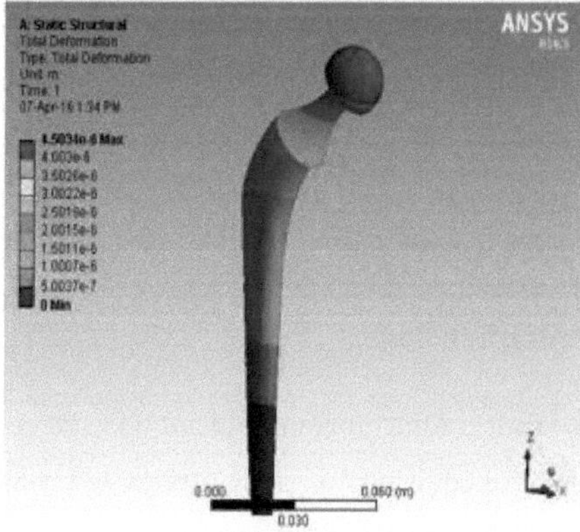

Figure 2- 3: Deformations

The results obtained from the finite element simulation of a human hip joint stem implant show that all six titanium alloys show very little variation. The highest strain value is presented by Ti-15Mo and the lowest by Ti-6Al-7Nb. Ti-12Mo-6Zr-2Fe has the highest weight and Ti-6Al-4V is the lightest of all materials.

2.3 Finite element modeling of the static load of a hip implant

This article by Katarina Colica, Aleksandar Sedmakb, Aleksandar Grbovicb, Uros Tatica, Simon Sedmaka, Branislav Djordjevica [29] presents a numerical study of a replacement implant for partial hip arthroplasty. The long-term stability of hip prostheses depends on the loads exerted on the joint. The forces occurring in vivo can be much greater than the recommended test values, as a typical walking cycle generates forces up to 6-7 times the body weight in the hip joint.

2.3.1 Modeling with MEF

The model was configured with suitable boundary conditions, including fixation of the lower implant surface along all degrees of freedom, and the load was applied in the appropriate direction relative to the apex of the prosthesis femoral head. FEM analysis of the prosthesis was carried out using ABAQUS software (Dassault Systèmes), to simulate slow walking on a flat surface for a Ti-6Al-4V alloy model.

Stresses were calculated in order to estimate the probability of prosthesis failure under the maximum 6 KN loads that can occur during a gait cycle. The Von Mises stresses on the stems, caused by the static analysis, are presented in the following figure:

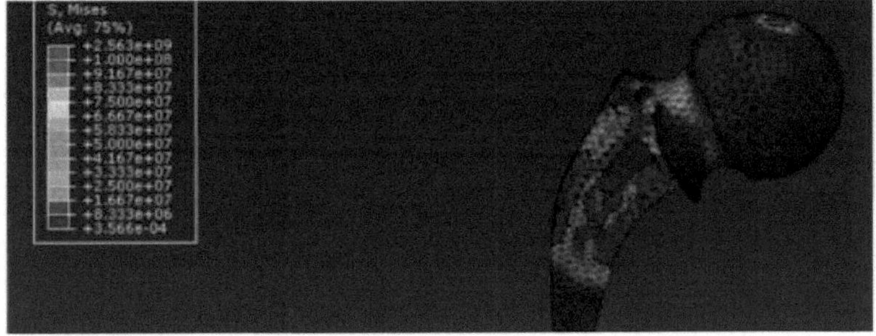

2.3.2   Results and discussion

Figure 2-5 shows a graphical comparison of the implant's displacement at the end of the calculation, under a maximum load of 6 kN, compared with the initial, undeformed state.

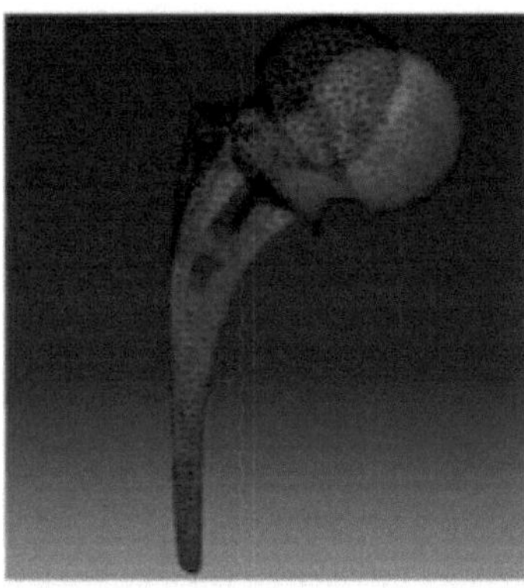

Figure 2- 5: Comparative representation of the implant's initial and maximum load states

The results of the numerical analyses of the proposed models are presented in the table below.

Table 2- 1: Results of numerical analyses of proposed models

Problems	Maximum Von Mises stresses (MPa)
Problems related to walking slowly on a flat surface	256.3
	361.2
Stair climbing problems	535.5
Stumbling problems	312.6
Problems going down the stairs	

The calculated Von Mises stresses, presented in Table 2.1, are significantly lower than the yield stresses of Ti-6Al-4V (860 MPa). The cross-section of the implant with the hole represents the location of the critical stress concentration. It is also the exact area in which fracture occurred in the specimens during operation (Figure 2-6).

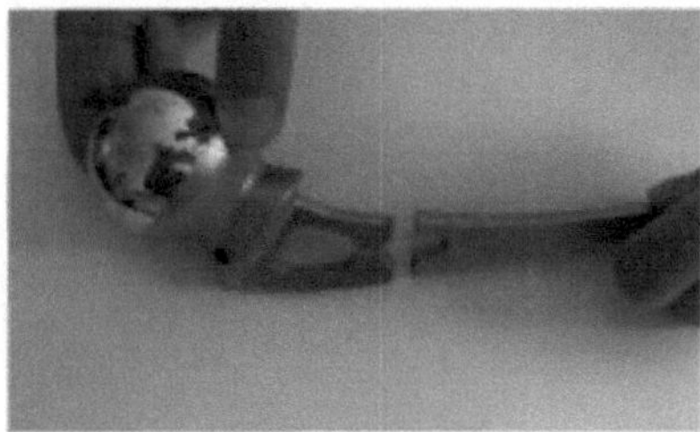

Figure 2- 6: Fracture of an artificial hip prosthesis

Figure 2-7 shows the total displacement of the prosthesis under the effects of maximum load.

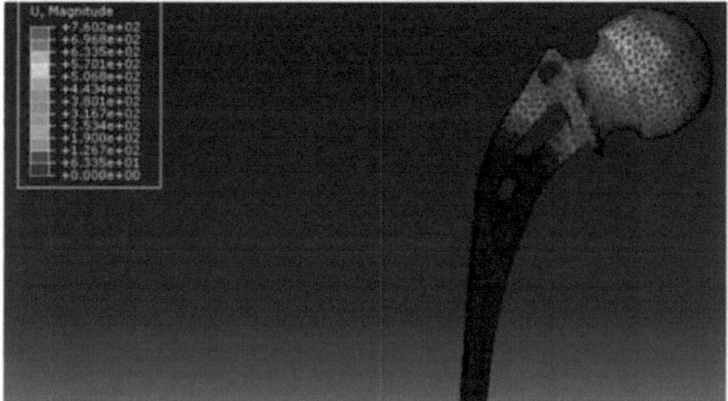

Figure 2- 7: Total displacement of the prosthesis under the effects of maximum load

2.3.3 Conclusion

For the results obtained from the static numerical analysis, the Von Misses stress values are below the yield strength of the Ti-6Al-4V and Co-Cr alloy, but above the yield strength of 316L steel, even under the highest assumed load magnitudes.

2.4 Numerical analysis of stress distribution on the artificial hip joint due to jumping activity

The main aim of this research, by Gilar Pandu Annanto, Eko Saputra, J Jamari, Athanasius Priharyoto Bayuseno, Rifky Ismail, Mohammad Tauviqirrahman and Iwan Budiwan Anwar [34], is to guarantee the safety of hip prosthesis using FEM. The prosthesis is subjected to the load that occurs from jumping activity. 3 materials were used in this research: titanium alloy (Ti-6Al-4V), cobalt-chromium alloy, and stainless steel (SS316L).

2.4.1 MEF modeling and simulation

2.4.1.1 Geometric conditions of the model :

The diameter and length of the femoral neck are 12 mm and 23.5 mm respectively. The stem has a diameter of 11 mm and a height of 142 mm. The bone cement layer is approximately 4 mm thick. Figure 2-8 shows the geometric model used in this study. The mesh used in the study is tetrahedral.

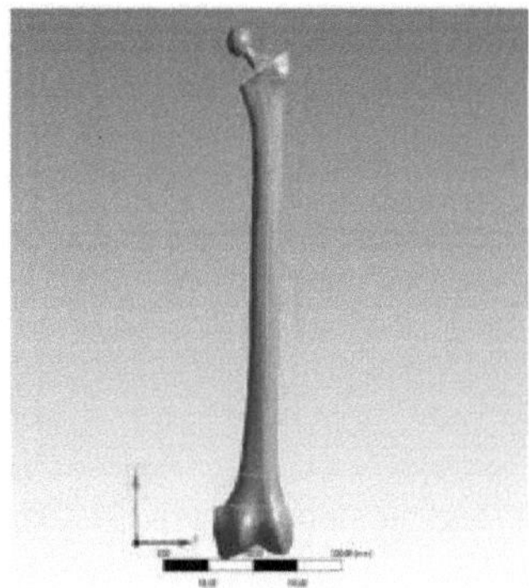

Figure 2- 8: Geometric modeling of the hip prosthesis

2.4.1.2 Boundary conditions

Determination of the loading model is based on the work of Senalp et al. In the dynamic analysis, F dynamics is applied to the surface of the femoral head. F dynamics represents the load produced by the body weight during a jumping activity. The load applied to the proximal zone of the greater trochanter is 1.25 kN at an angle of 20°. The bottom of the stem is loaded with 250N. Figure 2-9 shows the direction of the load applied to the prosthesis assembly with cemented bone.

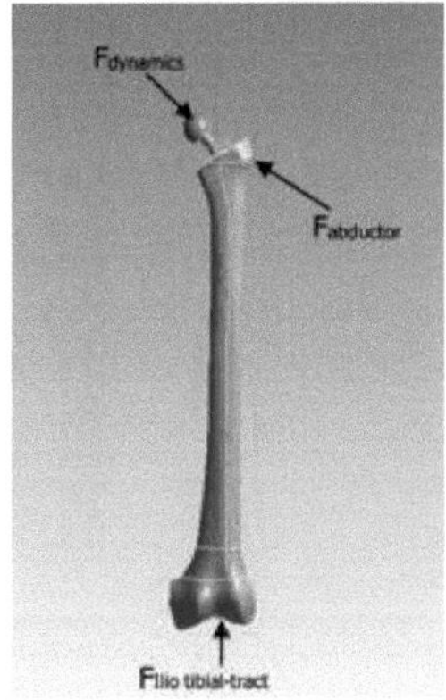

Figure 2- 9: Modeling the loads applied to the hip prosthesis assembled with the femur bone.

The dynamic forces applied in this study are those that occur during the jumping activity.

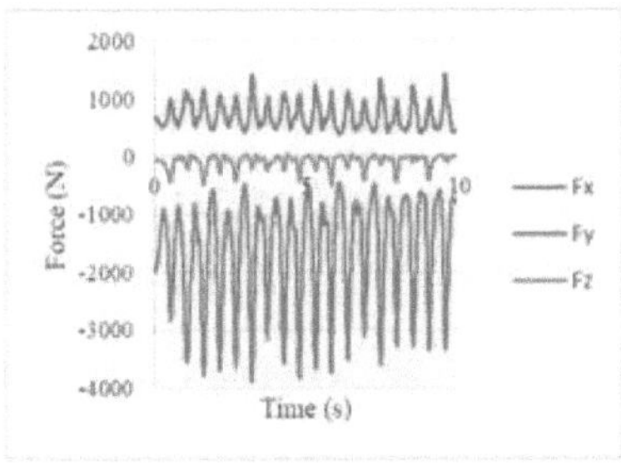

Figure 2-1: Dynamic forces applied to the hip during jumping activity

2.4.2 Results and discussion

Loads are applied to the hip prosthesis using 3 materials: Ti-6AL-4V, Co-Cr, and SS 316L.

Figure 2-11 shows the distribution of stresses on the hip prosthesis produced during jumping activity on Ti-6AL-4V.

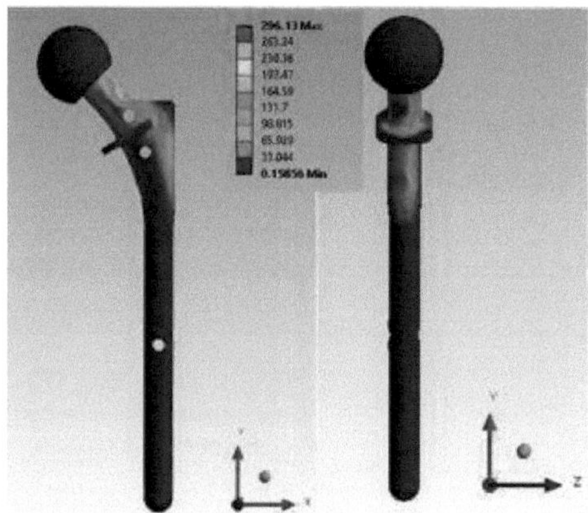

Figure 2-2: Distribution of stresses on the hip prosthesis produced during jumping activity on Ti-6AL-4V

The maximum stresses produced with Ti-6Al-4V, Co-Cr alloy and SS316L are 296.13 MPa, 294.02 MPa and 294.51 MPa respectively. Ti-6Al-4V outperforms the others with a safety factor of 2.7.

2.5 Hip joint simulator

Hip simulator studies have become an effective tool for basic research as well as pre-clinical testing to minimize the risks to patients when they receive new types of implants. The history of simulator development, mainly driven by research, has led to the development of many different models.

The aim of wear assessment is to determine the rate of wear and its dependence on test conditions, which include load, speed, temperature and spatial configuration of moving components. In order to obtain realistic results, a fatigue test must be carried out in such a way as to mimic working conditions as closely as possible.

The hip simulator mimics the load and multi-axial movements associated with activities of daily living, such as going up or down stairs, squatting to pick something up and running [35].

Figure 2-3: Hip joint simulator [35]

We can propose a design for a hip simulator test bench. The design is related to the simulation of human movement, as it is complicated for femoral heads, due to the angles of flexion and extension, adduction and abduction, inward and outward rotation.

Figure 2-13 shows an ISO 14242 hip joint simulator design by Nikolaos I. Galanis, Member, IAENG and Dimolrios E. Manolakos. Galanis, Member, IAENG and Dimitrios E. Manolakos [36].

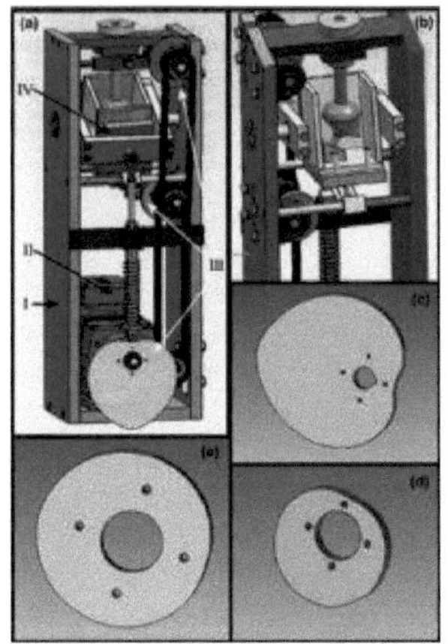

Figure 2-4: Design of a hip joint simulator [36]

With :

(a) Hip joint simulator. The entire machine is divided into (I) main body, (II) electric motor, (III) eccentric pulleys and (IV) reservoir. (b) A section of the reservoir, (c) eccentric extension and flexion pulley, (d) eccentric abduction and adduction pulley and (e) eccentric inward and outward rotation pulley.

Figure 2-14 shows other SolidWorks simulator designs.

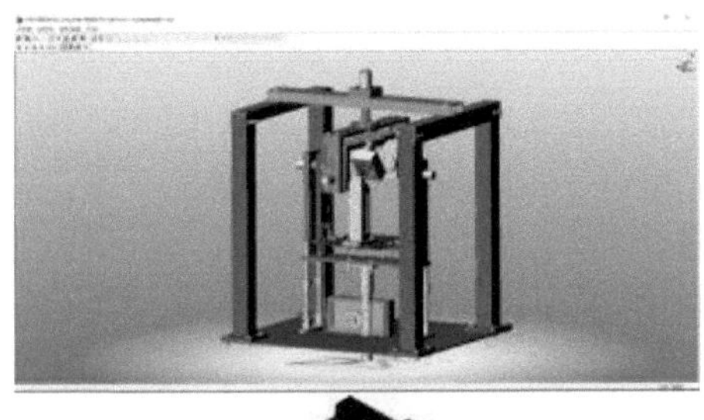

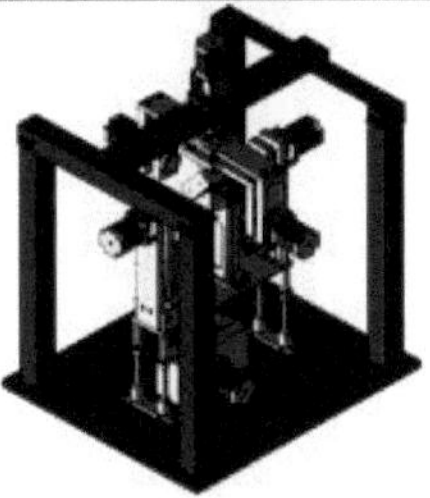

Figure 2-5: Design of a hip joint simulator test bench [37]

The results obtained from finite element models of the hip prosthesis are presented in the following table.

Table 2- 2: Summary of finite element models of hip prosthesis function

Authors	Software used	Applied load	Results	
			Constraint	Moving
Abhinav Kumar et al.	ANSYS	750 Pa	0.58 MPa	0.004 mm
Katarina Colica et al.	ABAQUS	6000N	535.5 MPa	0.76 mm
Gilar Pandu Annanto et al.	ANSYS	1250 N	296.13 MPa	–

2.6 Project orientation

The aim of this project is to simulate the load exerted on a hip prosthesis, made of Ti-6Al-4V alloy, as it occurs in the human body, in order to determine its strength, fatigue life and potential failure location.

Unlike previous studies, this one uses cyclic rather than static loading.

2.7 Conclusion

From this literature review, we can conclude that the Ti-6Al-4V used for hip prostheses can withstand high applied loads without risk of fracture. In the next chapter, we will carry out a numerical analysis using FEM on a real model of a hip prosthesis subjected to maximum loads.

CHAPITRE 3: NUMERICAL SIMULATIONS AND ANALYSIS USING THE FINITE ELEMENT METHOD (FEM)

3.1 Introduction

A prosthesis is the result of a design and manufacturing cycle whose initial characteristics are determined by the geometric conditions of the parts, the contact surfaces and the mechanical properties of the materials.

The finite element method is considered an indispensable tool for analysis and design. In recent years, the use of numerical simulations has expanded thanks to the improved performance of computing resources and calculation codes (ABAQUS, ANSYS, etc.). This makes it possible to model complex geometries. This method offers interesting prospects compared with analytical models.

A Muller-type Ti-6Al-4V titanium alloy hip prosthesis is proposed with two types of fixation: The 1er type is total fixation of the femoral stem: For verification of the strength of the neck, according to ISO 7206-6 (fatigue test for the neck) and The 2ème type is partial fixation: For verification of the strength of the stem, according to ISO 7206-4 (fatigue test for the stem).

3.2 Forces applied to the hip

This information is needed to test and improve the wear, strength and fixation stability of hip prostheses.

The hip contact forces in patients with different weights for one gait cycle are shown in figure 3-1 [38], which shows that the maximum force is 3000N according to ISO 14242-1.

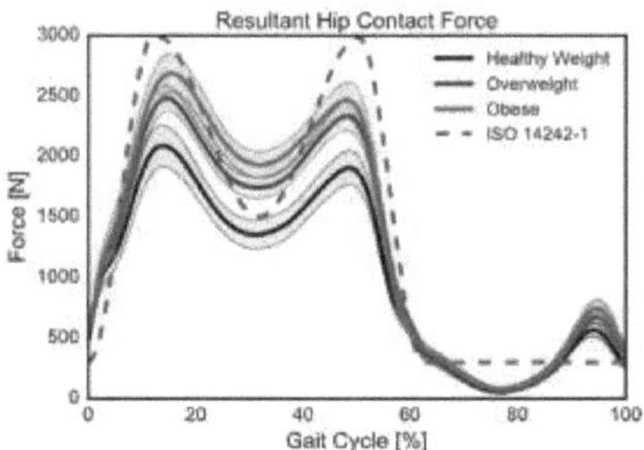

Figure 3- 1: Hip contact force in patients with different weights for one gait cycle

3.2.1 Hip contact forces during daily activities

The contact forces in the hip joint need to be known for strength, fixation, wear and friction testing of implants, to be able to model their designs and materials by computer simulation, and to give indications of activities that patients should avoid after prosthesis surgery.

Experimental observations by G. Bergman et al (2001) [39] show that the hip joint load of an ordinary patient is 238%P (of body weight) when walking at around 4 km/h, and slightly less when standing on one leg. The joint contact force is 251%P when climbing stairs and less than 260%P when descending.

The contact force of the hip with the magnitude F is transmitted by the acetabular cup to the implant head.

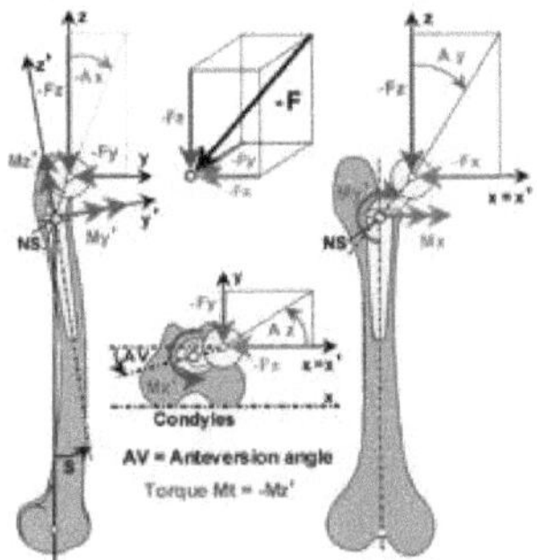

Figure 3- 2: Coordinate system for hip contact forces with F the hip contact force vector.

The patient goes through several activities during the day that apply contact forces on the hip Figure 3-3 illustrates 9 activities:

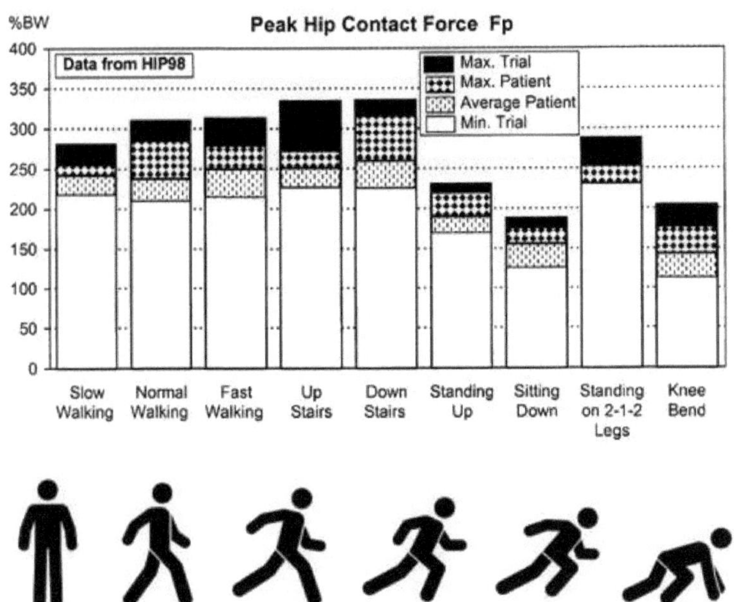

Figure 3- 3: Contact force values for each activity

The highest individual average of all patients (max. patient) can reach 340%P during stair descent (figure 3-3).

The results obtained (table 3.1) show that the max force of an ordinary patient is recorded during stair descent at 260%P.

Table 3- 1: Hip contact forces for each pattern of routine activities for an ordinary patient

	Force (%p)	Cycle times (s)
Slow walk	242	1.25
Fast running	238	1.11
Normal operation	250	0.96
Stair climbing	251	1.59
Stair descent	260	1.46
Getting up from a chair	190	2.49
Sitting on a chair	156	3.72
Switch to 2-1-2 leg	231	6.72

3.2.2 The load applied to the hip joint during walking and running

The force transmitted to the femoral head and prosthesis during walking and running was collected by Bergmann et al. 1993 [40] who, using instrumented femoral prostheses implanted in two elderly patients, determined the maximum load on the hip joint during walking and running.

3.2.2.1 The load applied to the hip joint during walking

Several tests were carried out at different speeds: the resultant force and its components were very similar for the left joint of patient 1 (Figure 3-4), the right joint of patient 1 (Figure 3-4(a)) and for patient 2 (Figure 3-4(b)). The peak force R_m increased with walking speed.

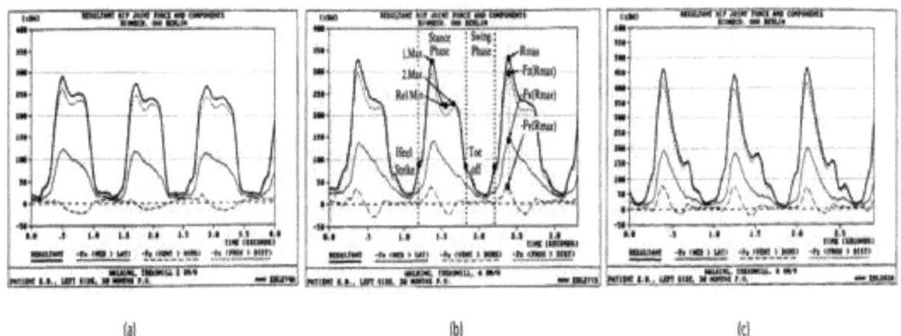

Figure 3- 4 Load results for patient 1's left hip joint while walking at speeds of (a) 2 Km/h, (b) 4 Km/h, (c) 6Km/H

The resulting force recorded during slow walking (2 km/h) is 280%P, 325%P during normal walking (4 km/h) and can exceed 450%P during fast walking (speed of 6 Km/h).

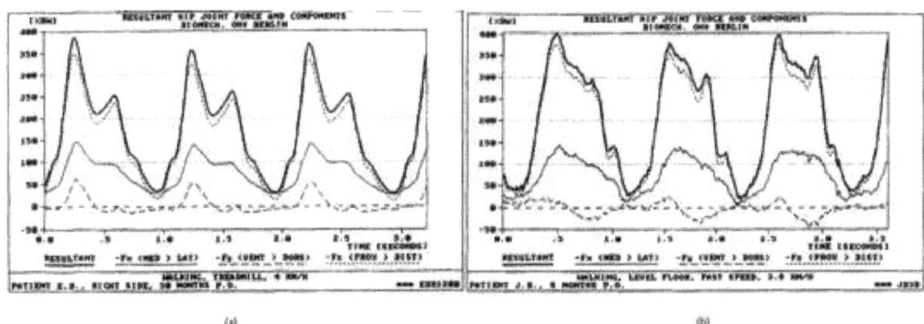

Figure 3- 5: Resultant force on the hip joint: (a) right hip of patient 1; (b) patient 2

The resulting maximum force recorded during normal walking (4 km/h) is almost the same for both patients: ±400%P.

3.2.2.2   The load applied to the hip joint during running

Figure 3-6 shows that the resulting force can reach up to 500%P during the stroke.

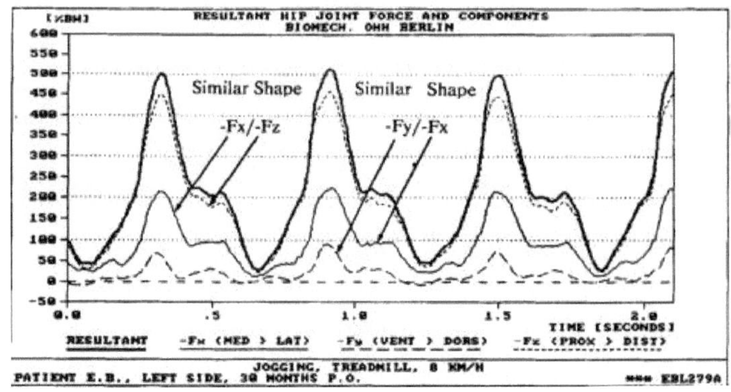

Figure 3- 6 : Result of the force applied to the hip at a speed of 8 km/h

3.3 Modeling and simulation

SolidWorks software is used to design the model of the hip prosthesis. Our model is a cemented Muller-type titanium alloy implant with the femoral head, diameter 38 mm, welded to the stem.

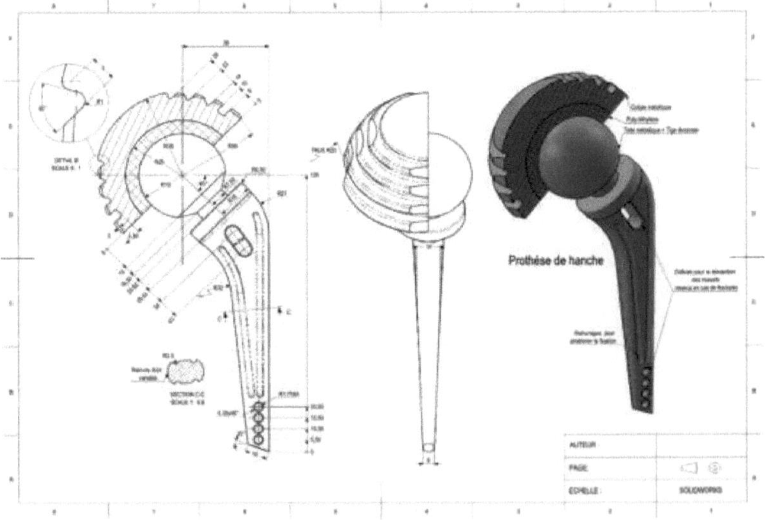

Figure 3- 7: Overall drawing of the hip prosthesis [Appendix 2].

And to analyze the loading results, we used the Ansys 2020 R2 support package.

The aim of this analysis is to determine the stress concentration points on the prosthesis.

The entire femoral stem and head are made of Ti-6Al-4V titanium superalloy. Loads are applied in accordance with Pauwels' work on pelvic loading during gait with monopodal support.

3.3.1 Material properties and behavior law

3.3.1.1 Material properties

The properties of the material used, taken from the Ansys technical data, are shown in the following table:

Table 3- 2: Properties of Ti-6Al-4V :

Density $(Kg/m)^3$	Young's modulus (MPa)	Modulus of elasticity (MPa)	Poisson's ratio	Yield strength (MPa)	Breaking strength (MPa)
4430	116000	114000	0.33	880	950

	A	B	C	D	E
1	Propriété	Valeur	Unité		
2	Variables des champs matériaux	Table			
3	Masse volumique	4508	kg m^-3		
4	Élasticité isotrope				
5	Dériver de	Module de Young e...			
6	Module de Young	1,1697E+11	Pa		
7	Coefficient de Poisson	0,33			
8	Module d'élasticité	1,1468E+11	Pa		
9	Module de cisaillement	4,3974E+10	Pa		
10	Courbe S-N	Tabulaire			
11	Interpolation	Log-log			
12	Echelle	1			
13	Décalage	0	Pa		
14	Limite élastique en traction	880	MPa		
15	Limite élastique en compression	880	MPa		
16	Limite à la rupture en traction	1144,5	MPa		
17	Limite à la rupture en compression	950	MPa		
18	nCode Multicurve Stress-Life Parameters				
19	Nfc	1E+30			
20	SEIs	0,15889			
21	Nm	1E+07			

3.3.1.2 Law of behavior

Behavior is said to be elastic and linear when, in a test, the stress-strain curve is the same at loading and at unloading, in which case the material's behavior is said to be elastic. Elastic behavior is linear when the strain tensor is proportional to the stress tensor during loading. The stress-strain relationship is linear, characterized by two parameters: an axial (Young's) modulus of elasticity E in the case of a simple compression or tension test, or a shear modulus G for a simple shear test, and Poisson's ratio v [41].

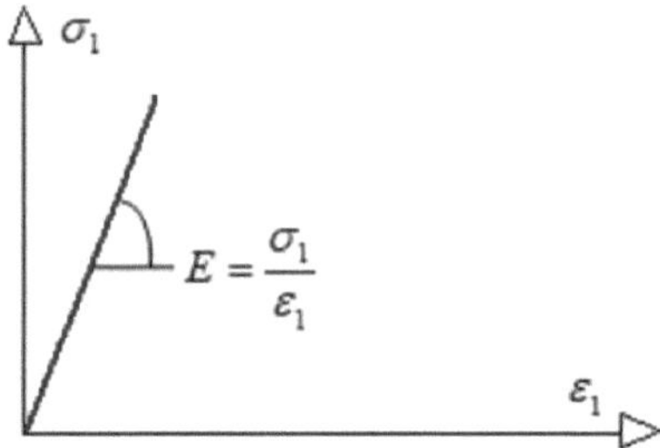

Figure 3- 9: Stress versus strain diagram

3.3.2 Model 1: Fully cemented femoral stem

3.3.2.1 Boundary conditions

The first model is set up in accordance with ISO 7206-6 (neck fatigue test) (Figure 3-10) with the following boundary conditions: the femoral stem is fixed in such a way that all degrees of freedom are eliminated and the load is applied to the femoral head, in accordance with ISO 14242-1 (Figure 3-1), presented by a force of 3000N (Figure 3-11).

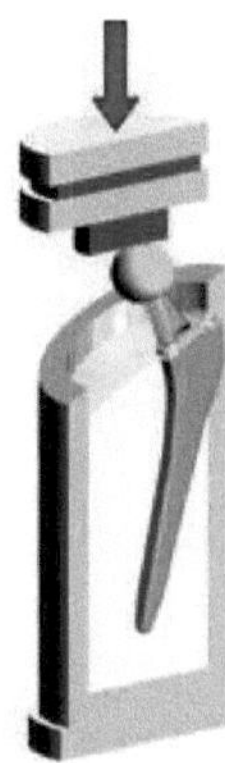

Figure 3-1: Fatigue test according to ISO 7206-6

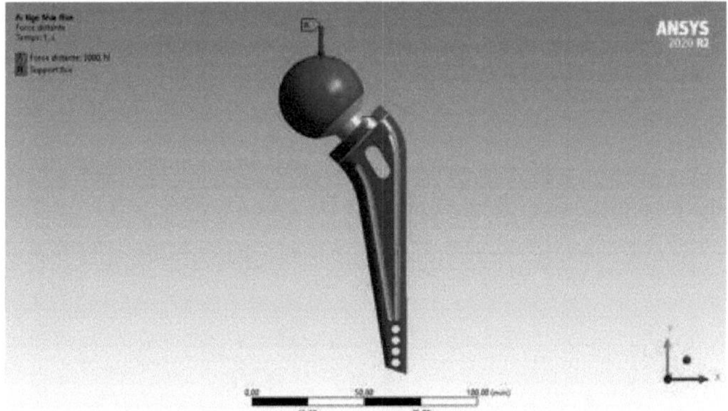

Figure 3-2: Boundary conditions applied to the hip prosthesis

3.3.2.2 Mesh associated with the model

The mesh elements chosen are tetrahedrons, which are best suited to meshing complex geometry.

For this model, we have chosen a mesh element size of 2 mm, as this is the 1ère size where the maximum stress value begins to stabilize within an optimal time, with refinement in the neck. The model is shown in the following figure:

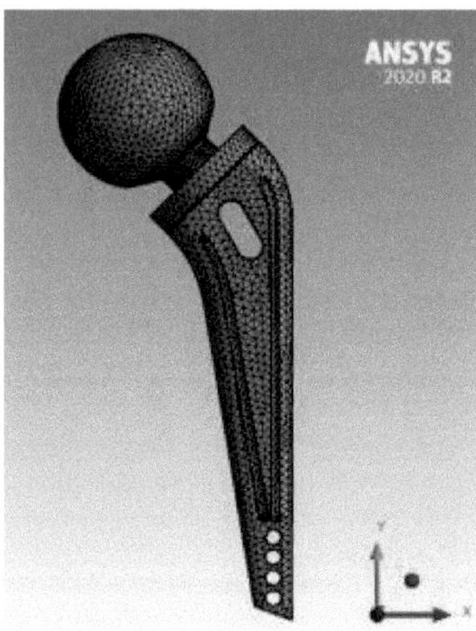

Figure 3-3: Mesh associated with model 1

3.3.2.3   The results

The Von Mises stresses resulting from the static analysis are shown in Figure 3-13, and the total displacement in Figure 3-15.

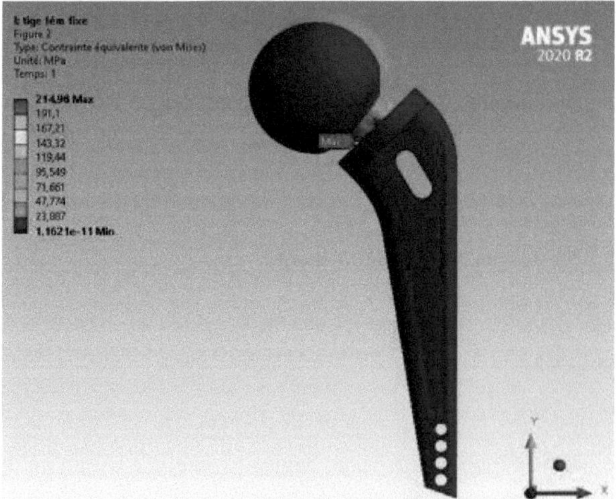

Figure 3-4: Von Mises equivalent constraints

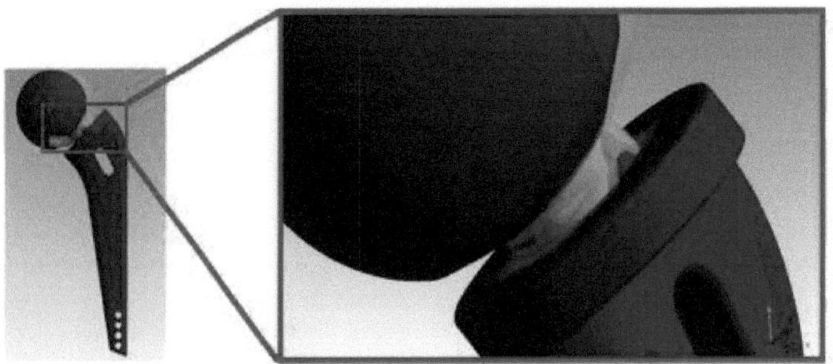

Figure 3-5: Predicted fracture zone

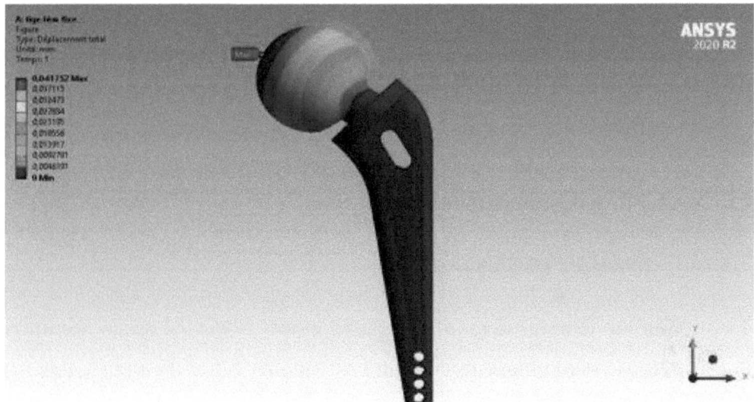

Figure 3-6: Total displacement

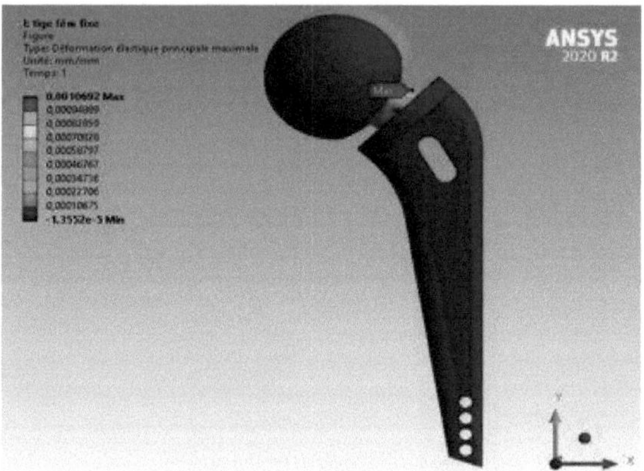

Figure 3-7: Maximum elastic deformation

It can be seen from the finite element simulation that the hip prosthesis resists the applied loads with max stress 215 MPa < σ_e = 880 MPa and max displacement 41µm.

3.3.3 Model 2: Semi-cemented femoral stem

3.3.3.1 Boundary conditions

Fixation of the femoral stem in the second model is carried out in accordance with ISO 7206-4 (stem fatigue test) (figure 3-17): only 2/3 of the stem is fixed and the load is applied to the femoral head and presented by a force of 3000N (figure 3-18).

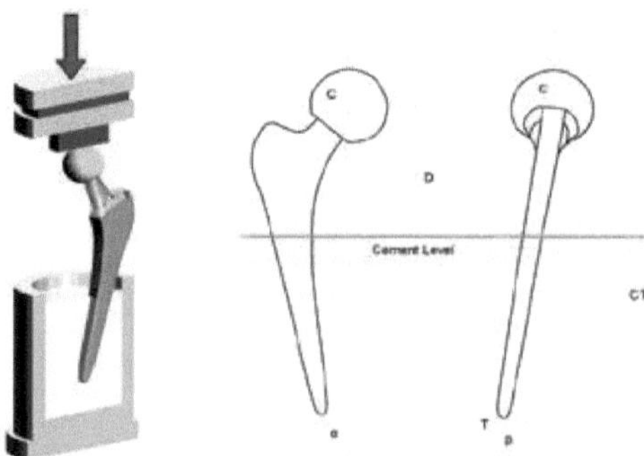

Figure 3-8: ISO 7206-4 fatigue test

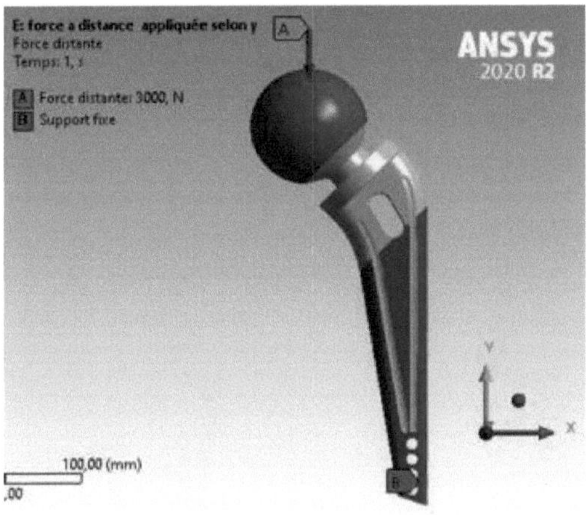

Figure 3-9: Force applied to the hip prosthesis

3.3.3.2   Mesh associated with the model

For this model we have chosen the same mesh as model 1 with a refinement in the uncemented area of the stem, presented in the following figure:

Figure 3-10: Mesh associated with model 2

3.3.3.3 The results

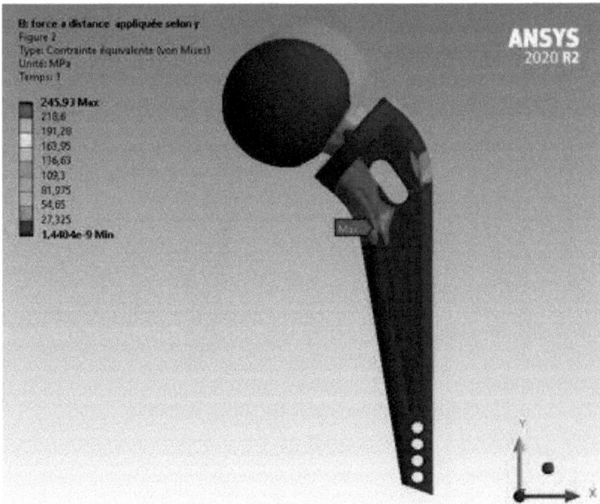

Figure 3-11: Von Mises equivalent stress

Figure 3-12: Predicted rupture zone

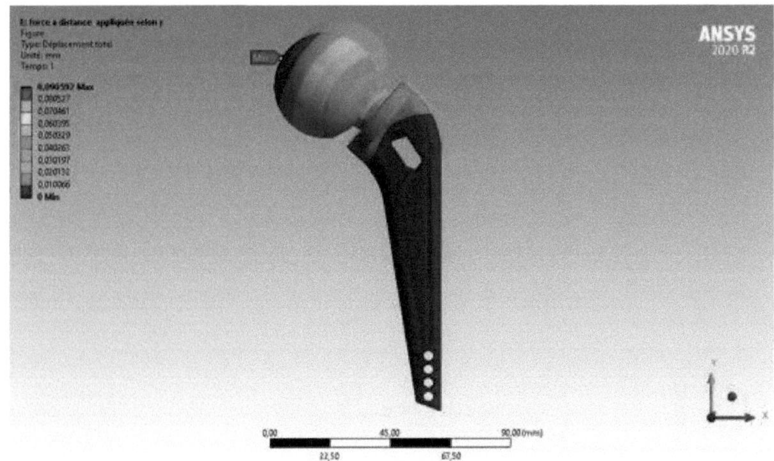

Figure 3-13: Total displacement

According to the finite element simulation, the hip prosthesis resists the applied loads with a max stress of 246 MPa $<$ σ_e = 880 MPa and a max displacement of 90 µm. Figure 3-19 shows the emergence of a new plane where failure is likely to occur in the semi-cemented model.

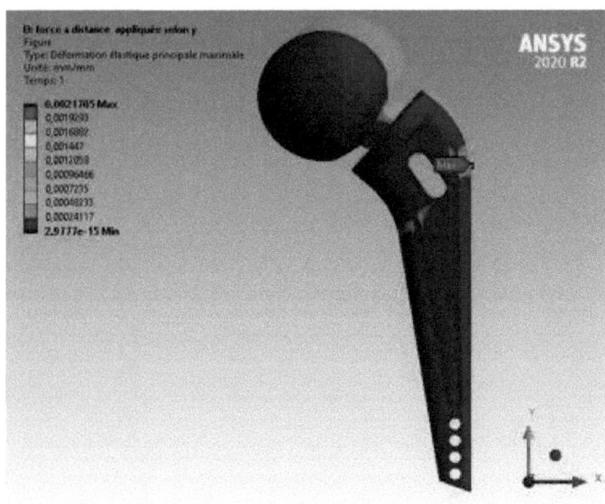

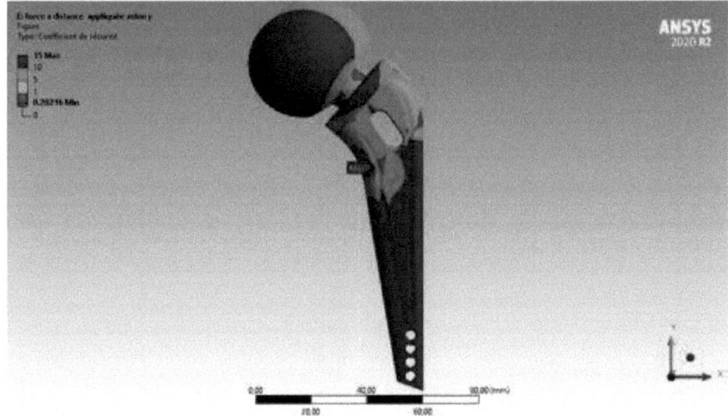

Figure 3-14: Main elastic deformation

Figure 3-15: Safety factor

Based on the test carried out by G. Bergmann et al. the resulting force can reach up to 500%P during the stroke. Assuming a person weighing 120kg, the force applied is 6000N (120x9.8x5). That's why we've gone back to work with the previous models, but with a maximum force of 6000 N, which is twice the value required by IS0 7206-7.

- Model 1:

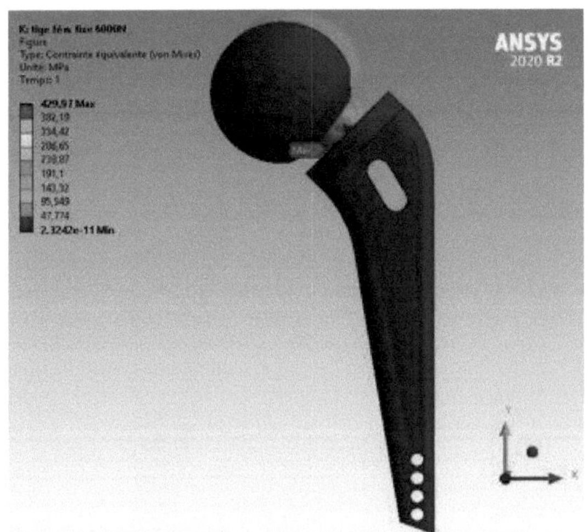

Figure 3-16: Von Mises equivalent stress

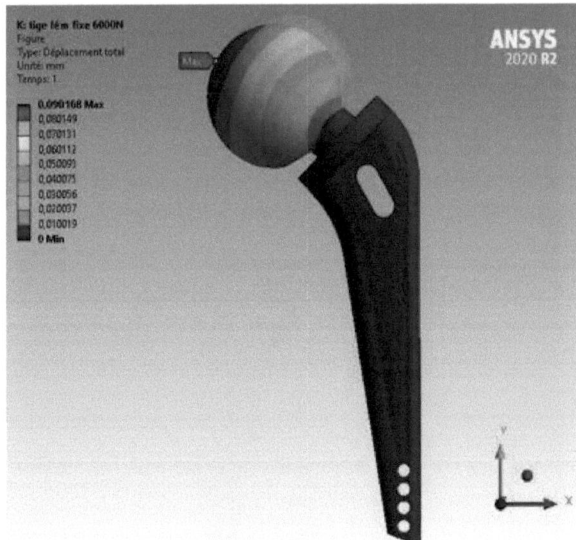

Figure 3-17: Total displacement

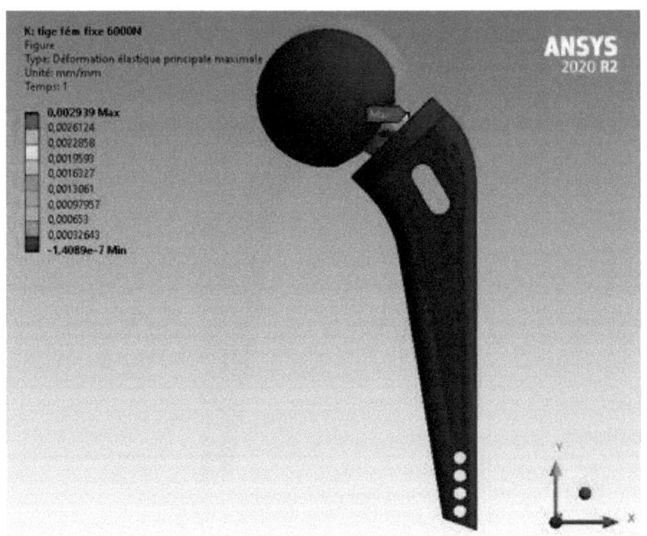

Figure 3-18: Maximum elastic deformation

Safety coefficient :

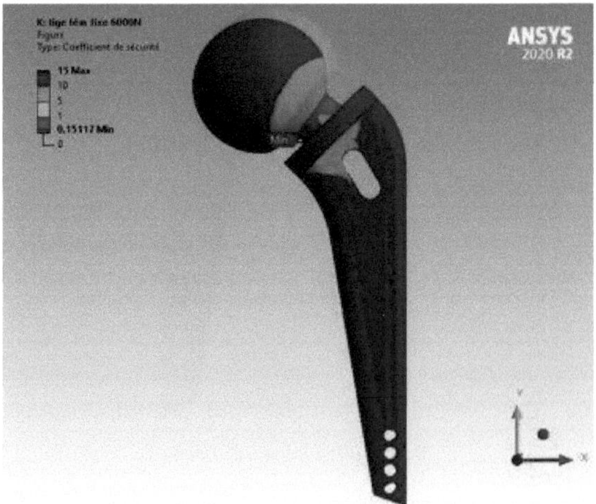

Figure 3-19: Safety factor

- Model 2:

C: max force a distance appliquée selon y
Force distante
Temps: 1. s
A Force distante: 6000, N
B Support fixe
ANSYS
2020 R2
Y
X

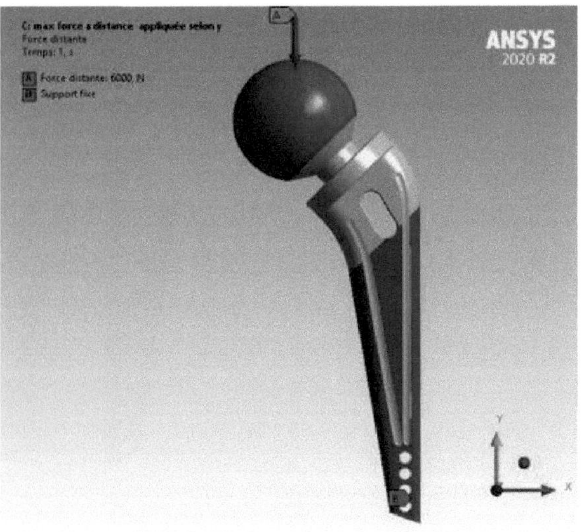

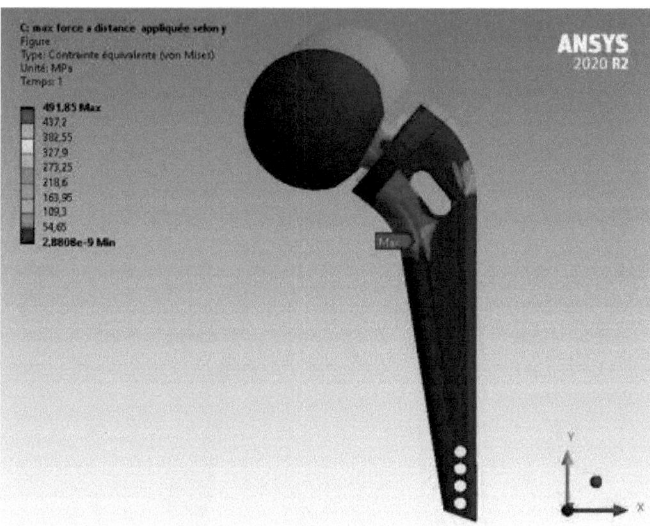

Figure 3-21: Von Mises equivalent stress

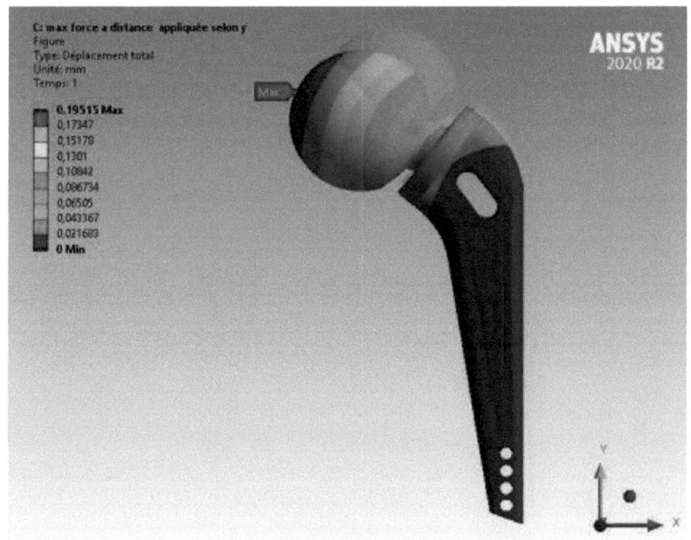

Figure 3-22: Total displacement

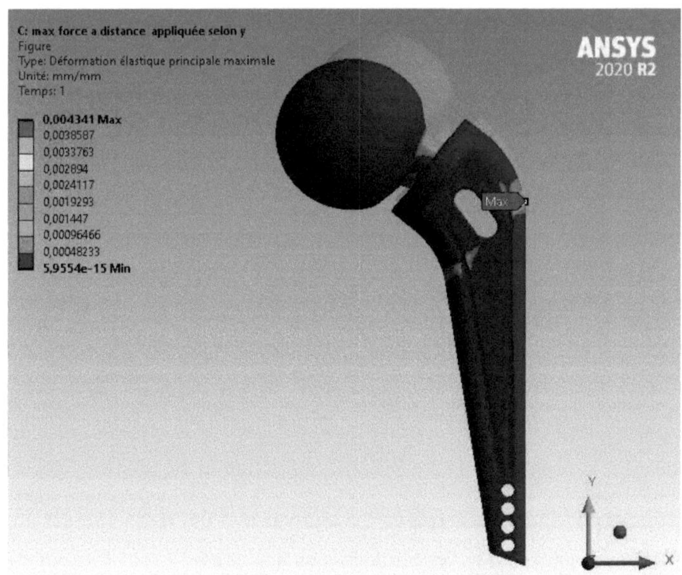

Figure 3-23: Maximum elastic deformation

87

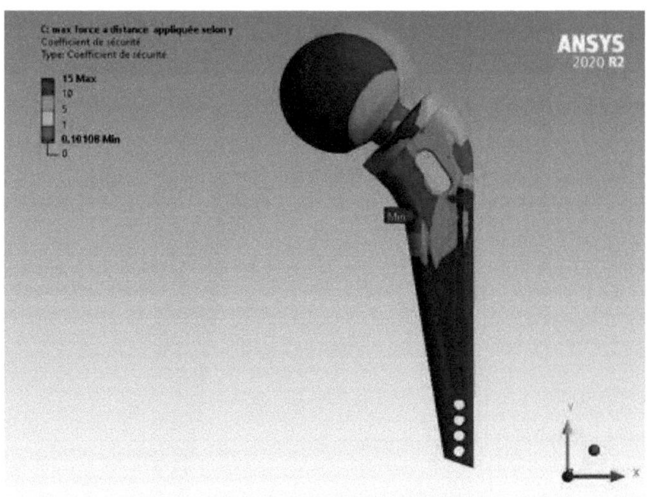

Figure 3-24: Safety factor

The results obtained from the numerical analysis are illustrated in the following table:

Table 3- 3: Maximum stresses with different forces for the two hip prosthesis models

Model	Force (N)	σ_{Max} (MPa)	Max. elastic deformation (mm/mm)	Total displacement (mm)
1	3000	214.98	0.001069	0.04175
	6000	429.97	0.002939	0.09016
2	3000	245.93	0.002170	0.09059
	6000	491.85	0.004341	0.19515

The results show that the distribution of equivalent Von Mises stresses along the implant is non-uniform, with the highest stress marked in the proximal zone at the point of contact with the neck of the femoral stem for the 1er model and in the 2ème model near the oblong hole. There is a risk of fracture in the proximal zone.

In the posterior and anterior parts, the equivalent Von Mises stresses are zero, so the implant is subjected to very little mechanical load.

If we compare the σ_{Max} results obtained, 215 MPa from model 1 and 246 MPa from model 2, with the yield strength of the Ti-6Al-4V implant (880 MPa), we see that they are still within the acceptable safety range.

Even at a maximum load of 6000N, the σ results$_{Max}$, for models 1 and 2 are 430 MPa and 492 MPa respectively, remain below 880 MPa.

3.4 Conclusion

The objective of hip prosthesis design is to have low stress, low displacement and low wear with a very high fatigue life. It can be concluded that the application of the finite element method (FEM) is a good alternative approach to provide preliminary results and insight into the mechanical properties of potential implant models.

CHAPITRE 4: STUDY AND SIMULATION OF THE FATIGUE BEHAVIOR OF HIP PROSTHESIS

4.1 Introduction

To determine the strength of the Ti-6Al-4V hip prosthesis, the fatigue life of the prosthesis and the location of potential failure, we'll run a simulation using nCode DesignLife software. Motion loading is assumed to be cyclic, with a periodic sinusoidal fatigue cycle.

4.2 General information on fatigue

The term fatigue generally refers to the slow degradation of materials due to the application of repeated stress cycles over time. In fatigue, materials are characterized by tests in which a specimen is subjected to alternating sinusoidal loading of constant amplitude until crack initiation is observed.

The number of cycles to initiation obtained experimentally is then compared with the number of cycles to failure of the structure. Such tests are repeated for different levels of loading amplitude to establish the material's Wöhler curve, giving on the ordinate the stress amplitude noted σ_a or σ (depending on the loading ratio R) as a function of the cycle at initiation or at failure (in logarithmic scale).

Fatigue is characterized by the endurance limit σ_D , which represents the stress level at which crack initiation (fracture) can occur. The probability of failure depends on the number of cycles N, load ratio R and temperature (frequency is not important for metallic materials) [42].

4.2.1 Definitions

Fatigue is a mode of failure that occurs when a material is subjected to the action of cyclic, repeated or alternating stresses or strains. This action can lead to changes in a material's local properties, resulting in crack formation and eventual structural failure.

The danger of fatigue exists in two cases: when the loading level is below the elastic limit, or when cracking is not visible to the naked eye.

The main stages of fatigue are :

Crack initiation, crack propagation and final fracture.

4.2.2   Fatigue stress cycles

Each time there is a variable stress cycle over time, there is a risk of rupture. The actual stress cycle is random, but in laboratories the stress cycle is simplified to a periodic cycle.

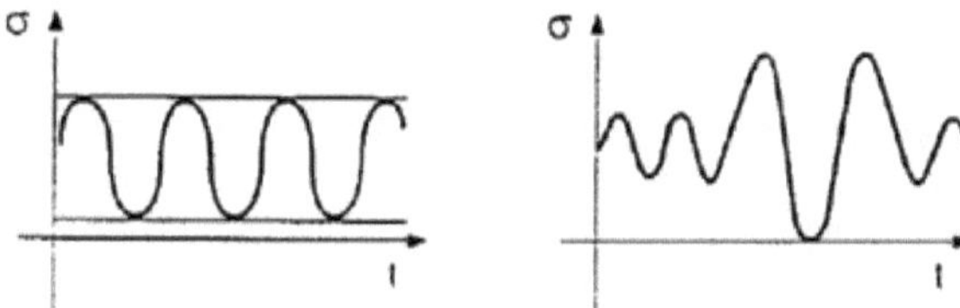

Figure 4- 1: Fatigue cycle: (a) periodic effort cycle, (b) random effort cycle

    (a)   (b)

The stress cycle is the periodic repetition of the stress-time function. Sinusoidal stress can be considered as the superposition of an alternating stress σ_a and a static stress σ_m called average stress.

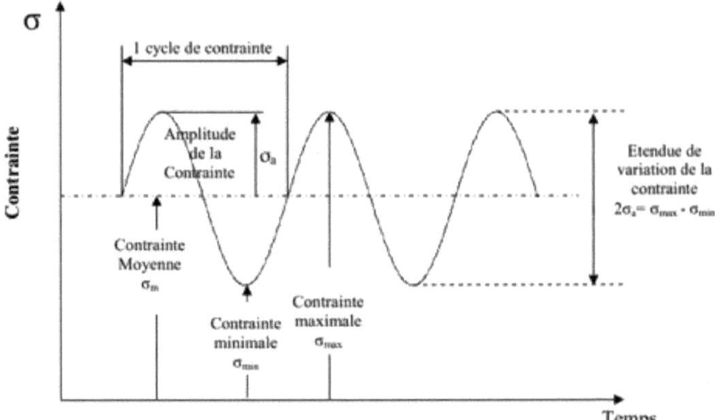

Figure 4- 2: Fatigue stress cycle

The characteristics of a fatigue cycle are shown in Figure 4-2 :

Stress amplitude: $\sigma_a = (\sigma_{max} - \sigma_{min})/2$

Extent of stress variation : $\Delta\sigma = \sigma_{max} - \sigma_{min}$

Maximum stress: σ_{max}

Minimum stress: σ_{min}

Average stress: $\sigma_m = (\sigma_{max} + \sigma_{min})/2$

Load ratio: $R = \sigma_{min} / \sigma_{max}$

Fatigue cycles depend on load ratio, average stress and stress modulus.

The possible forms of the constraint cycle are shown in the following figure:

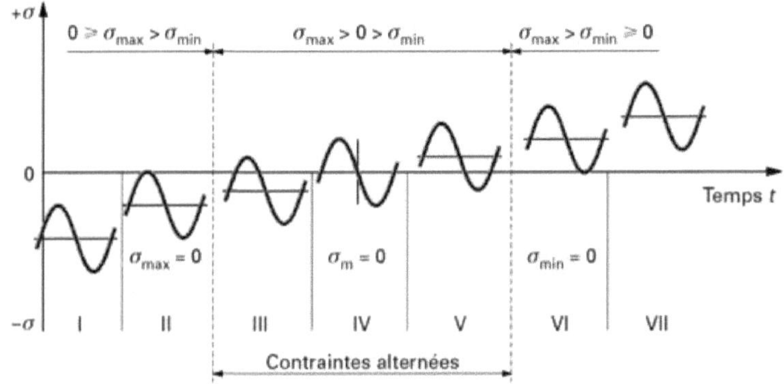

Figure 4- 3: Stress cycle shapes

Table 4- 1: Types of constraint

Type of constraint	Compression		Traction	
	Zone	Load ratio	Zone	Load ratio
Corrugated	I	$1 < R_\sigma < +\infty$	VII	$0 < R_\sigma < 1$
Repeated	II	$R_\sigma = +\infty$	VI	$R_\sigma = 0$
Alternating asymmetrical	III	$-\infty < R_\sigma < -1$	V	$-1 < R_\sigma < 0$
Purely alternating	IV	$R_\sigma = -1$		

4.2.3   Fatigue strength of materials

To apply a fatigue test to a mechanical part, you need to know the material's yield strength Re or $R_{p0.2\%}$, , the breaking strength Rm (to calculate resistance to static loading) and the material's mechanical characteristics.

4.3 Fatigue laws

There are various methods for predicting fatigue life in the presence of medium stresses.

In Figure 4-4, the Goodman, Gerber and Soderberg methods are shown in the Haigh diagram.

Goodman's equations:

$\sigma_a = \sigma_D .(1 - \sigma /R)_{mm}$

Gerber equations:

$\sigma_a = \sigma_D .(1 - \sigma /R)_{mm}^2$

The Soderberg equation:

$\sigma_a = \sigma_D .(1 - \sigma /R)_{me}$

With σ_a the stress amplitude, σ_m the mean stress, σ_D the fatigue strength corresponding to the stress amplitude in the case $\sigma_m =0$, R_m the tensile strength and R_e the yield strength.

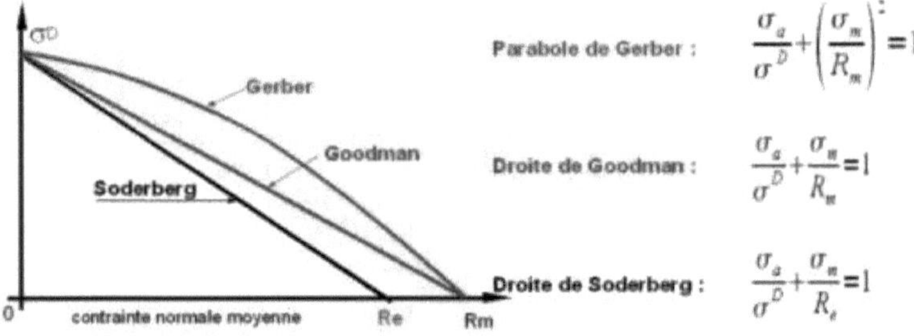

Figure 4- 4: Haigh diagram with representation of the empirical laws of Gerber, Goodman and Soderberg

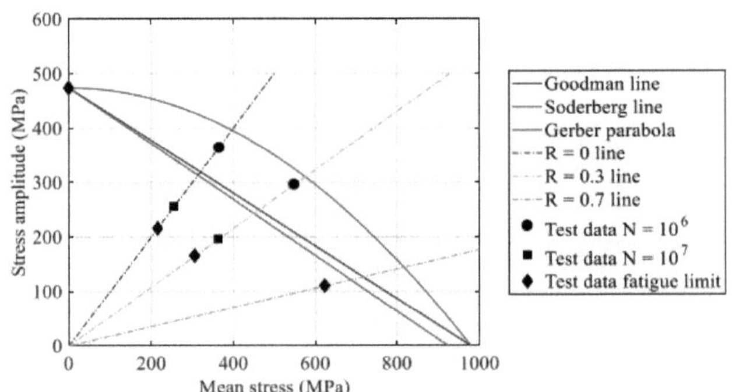

Figure 4- 5: Haigh diagram for Ti-6Al-4V [43]

- Goodman's Law

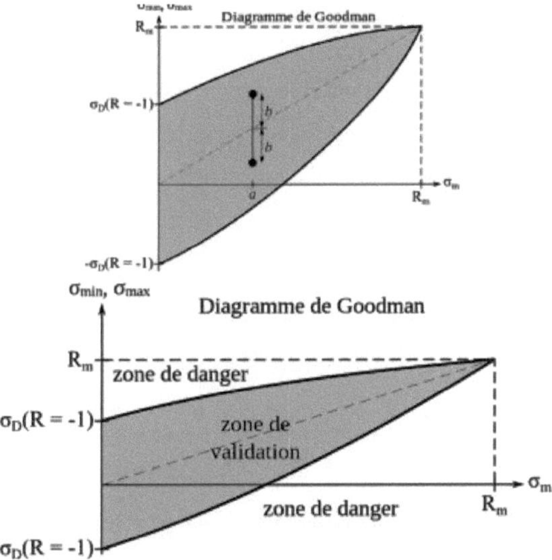

Figure 4- 6: Goodman diagram

The simplest form is Goodman's linear law:

$\sigma_D (R) = \sigma_D (R = -1) \cdot (1 - \sigma / R)_{mm}$

If we are in a case of $\sigma_m = a$ and $\sigma_a = b$

We place the segment [(a ; a - b) ; (a ; a + b)], if the segment is inside the validation zone, then we consider that the design is validated, that the part will resist.

If the point is outside the safety zone, we are in the danger zone, the zone of limited lifetime ($N_{rupture} < 10^7$); the design is not validated. [44]

4.4 Simulations

Numerical simulations of a fatigue test were created using ANSYS 2020R2 nCode DesignLife software to determine service life and damage, under cyclic loading of sinusoidal form at room temperature and loading ratio $\Delta\sigma=10$, for the stem of a Ti-6Al-4V Titanium alloy hip prosthesis.

4.4.1 Software description: nCode DesignLife

nCode DesignLife is a fatigue analysis system based on CAE (Computer Aided Engineering: a process of solving engineering problems through the use of complex graphical software), which contains a comprehensive set of solvers and advanced methods for predicting the durability of structures from key finite element analysis result files. It identifies critical locations and calculates realistic fatigue lifetimes for metals and composites.

4.4.1.1 Features

nCode is an advanced technology for analyzing multiaxial fatigue, welds, short-fiber composites, vibration fatigue, crack propagation and thermomechanical fatigue with an intuitive graphical interface for fatigue analysis from FEA results such as ANSYS, Nastran, Abaqus, Altair OptiStruct, LS-Dyna and others.

4.4.1.2 Benefits

- Reduce the need for physical testing and avoid costly design and tooling modifications
- Perform intelligent, rapid physical tests by simulating them first
- Reduce warranty claims by reducing failures
- Reduce cost and weight by testing several design options
- Improve consistency and quality with a standardized analysis flow
- Highly configurable for experienced users

4.4.1.3 nCode DesignLife options

Stress-Life (SN), Strain-Life (EN), Factor of Safety, Fatigue of Weld Points, Fatigue of Weld Beads, Vibratory Fatigue, Thermomechanical Fatigue (TMF), Short/Continuous Fiber Composites, Bonded Assemblies, Distributed/Multicore Analysis, ...

4.4.2 Ansys nCode methodology

Ansys nCode DesignLife is a tool for durability analysis, offering a complete fatigue diagnosis process for predicting product life.

This software works in parallel with Ansys Mechanical to reliably assess fatigue life. Ansys nCode can calculate the damage caused by cyclic loading to determine the expected life of a product, based on the results of finite element analysis from Ansys Mechanical and Ansys LS-DYNA.

The complete process of numerical simulation of the hip prosthesis stem and determination of fatigue life carried out by ANSYS products combines several simulation tools:

Firstly, we used ANSYS Mechanical, for the structural finite element analysis of our product, which simulates the stresses and strains resulting from a single loading.

Next, we moved on to ANSYS nCode DesignLife, which simulates the life of the rod by combining the results of finite element analysis with cyclic loading and material data.

4.4.3 Simulation description

ANSYS Workbench provides the environment for integrating nCode with other ANSYS structural modules, as shown in figure 4-7.

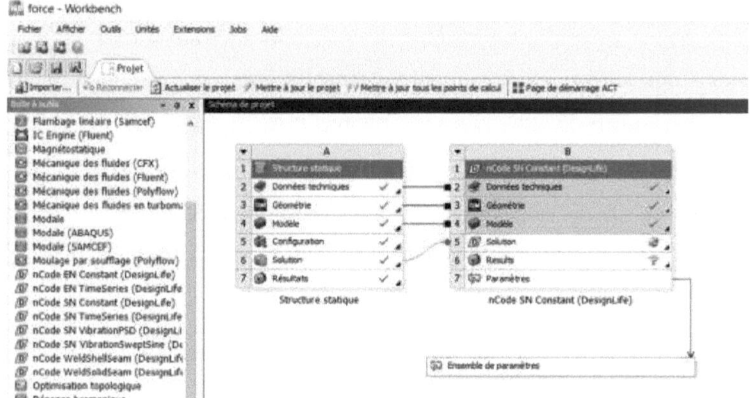

Figure 4- 7: nCode integration in Ansys Workbench

The aim of this simulation is to estimate the damage to a femoral stem subjected to high loads, in order to gain confidence and demonstrate the reliability of the device.

The nCode software works with so-called "glyphs" which can be adapted to a workflow depending on the problem to be solved. As can be seen in Figure 4-8, for our problem, the glyphs on the left enter simulation from Ansys mechanical, loading and material properties. All converge on the main solver, the stress analysis glyph.

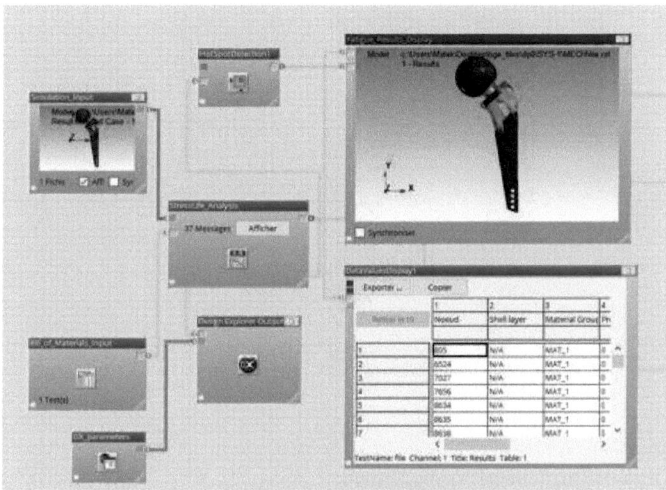

Figure 4- 8: nCode workflow

4.4.4 Results

- Model 1:

Once nCode has carried out the calculations, the results for damage and lifetime are displayed as shown in Figures 4-9 and 4-10.

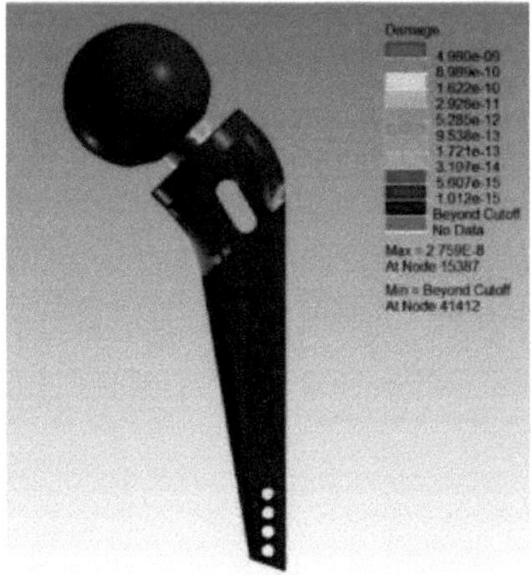

Figure 4- 9: Damage results for the fully cemented model

The maximum damage is calculated at the neck area, where the damage is 1.432e-08. If the damage were equal to unity, failure would be certain. Since the calculated damage is far less than 1, working conditions cannot damage the part.

Figure 4-1: Service life calculated by nCode for the fully cemented model

The minimum service life is calculated at 1,120e+07 also in the neck area.

- Model 2:

Damage and service life results are shown in Figures 4-11 and 4-12.

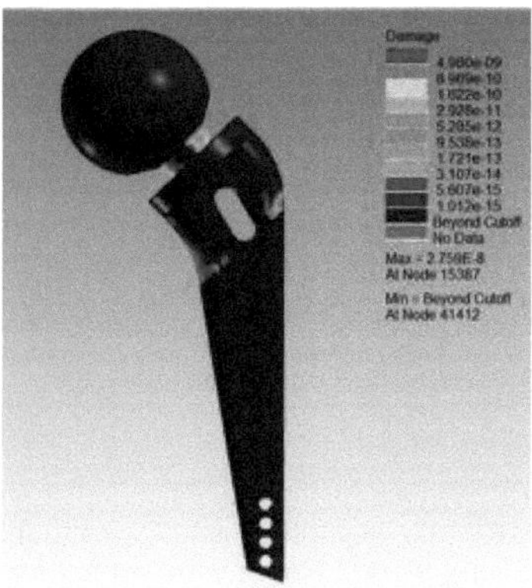

Figure 4-2: Illustration of damage in the semi-cemented model

The maximum damage is calculated at the neck area, where the damage is 4.978e-09. The prosthesis is far from damaged.

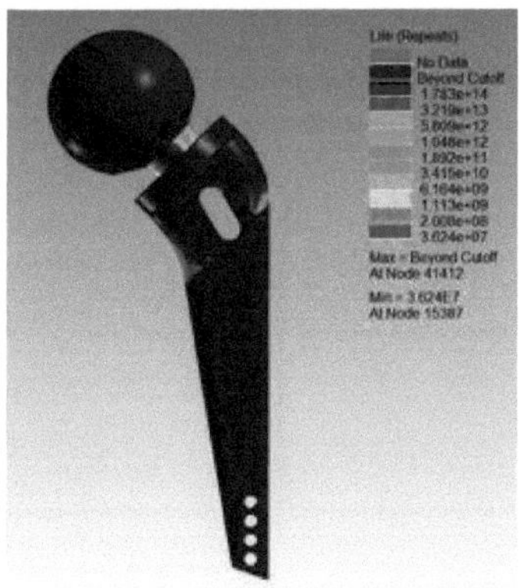

Figure 4-3: Service life for the semi-cemented model

The minimum lifetime in the neck area is 3.626e+07.

The results obtained for the two models with a force of 3000N are shown in the following table:

Table 4- 2: The results obtained for the two models with different

	Damage (Max)	Service life (Min)
Model 1	1.432e-08	1.120e+07

| Model 2 | 4.978e-09 | 3.626e+07 |

The neck of the prosthesis shows greater damage per cycle, and has a shorter lifespan compared to the stem. This suggests that the neck is the part of the prosthesis most likely to deteriorate first.

The stem of the prosthesis is more resistant to fatigue, with a lifespan three times longer than that of the neck, and less maximum damage.

4.5 Conclusion

From our simulation using nCode DesignLife under Ansys Workbench, we can conclude that the prosthesis can withstand long lifetimes. And if we want to extend the life of these prostheses, we propose to apply a post treatment to the critical areas illustrated in the previous simulations, such as superficial laser shock treatment (LSP).

CONCLUSIONS AND OUTLOOK

With the development of biomaterials used for medical implants and due to scientific progress, human life is becoming easier and more comfortable. In this context, a strength study was carried out on implantable stems for hip prostheses made from Titanium Ti-6Al-4V superalloy, with the aim of determining the strength of the femoral stem, the polycyclic fatigue life of the prosthesis and the location of potential failure.

We began this book with a bibliographical search illustrating various studies on the anatomical data of the hip, and synthetic studies on hip prostheses, including their history, compositions, stresses and strains of implantable stems, and the characteristics of the materials used, in order to gain a better understanding of how hip prostheses work.

After the bibliographic study required to introduce hip prostheses, we moved on to a review of the literature concerning the most relevant work that has been carried out.

The loads applied to hip prostheses in the course of human activities generate stresses that vary over time, which can lead to implant failure and material fatigue. In this context, the third chapter presents a numerical simulation using FEM to determine the maximum stresses, total displacements and elastic deformations due to the application of maximum loads on the femoral stem of a hip prosthesis.

Predicting the fatigue life of a hip prosthesis offers advantages for young sports patients.

In the final chapter, a numerical simulation was carried out using nCode DeseignLife software to determine the fatigue life of the prosthesis and the location of potential failure. It was assumed that the loading cycle of a movement is sinusoidal.

It has been assumed that the fatigue cycle is sinusoidal, whereas there is a difference between the results predicted by standard implant tests and the actual loads that may occur in practice. For this reason, we aim to analyze the prosthesis under loads corresponding to body weight, as well as under the maximum actual loading that is assumed to occur during the movement cycle (dynamic loading).

In this way, a practical fatigue test can be carried out on a real model to compare the results with the numerical analysis. It would also be interesting to tackle the research topic of the manufacture of medical prostheses printed by metal additive manufacturing.

APPENDICES 1

A1. **Hip surgery**

Total hip replacement requires access to the joint between the pelvis and the femur. Various routes are available to access this joint. The minimally invasive anterior approach has the advantage of preserving the hip's environment, without cutting into muscle or tendon.

Hueter's minimally invasive anterior approach offers numerous advantages. Unlike the techniques commonly used (Moore posterolateral approach, Hardinge anterior approach, Trans-trochanteric approach), this approach preserves the surrounding anatomical structures, since it allows access to the hip without musculotendinous or bony transection. [45]

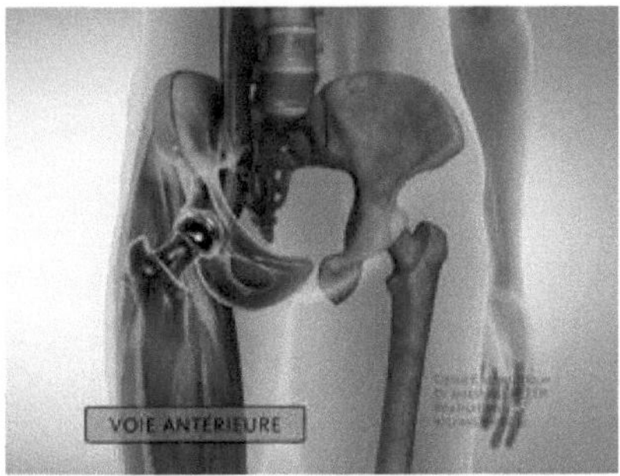

Figure A1- 1: Minimally invasive anterior total hip replacement

A1.1 Minimally invasive anterior approach

After a short opening in the skin, access to the joint is gained by progressively reclining the various muscle layers, using retractors, down to the capsule, the joint's envelope.

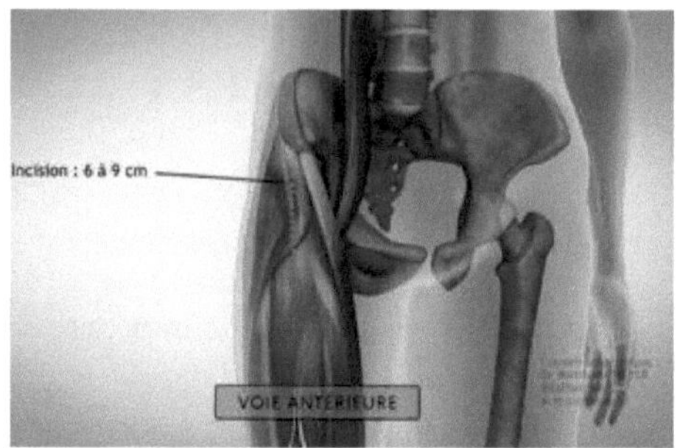

Figure A1- 2: 6 to 9 cm incision

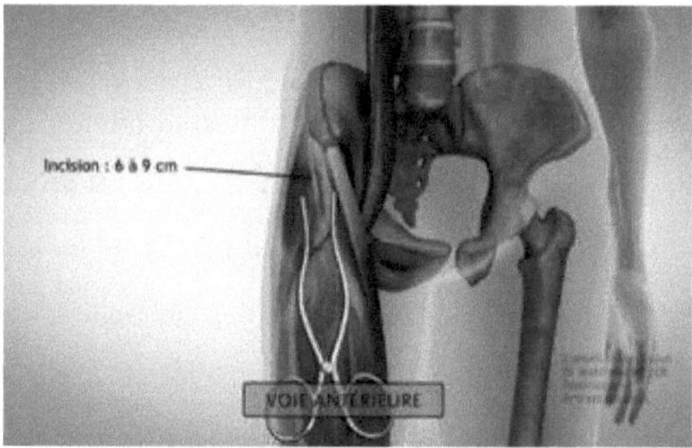

Figure A1- 3 : Skin opening with retractors

The joint capsule is opened to expose the femoral neck, which is sectioned with an oscillating saw and the femoral head removed.

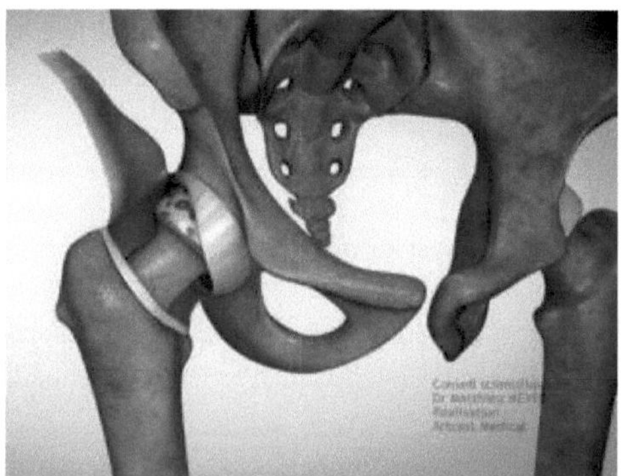

Figure A1- 4: Opening the capsule and exposing the femoral neck

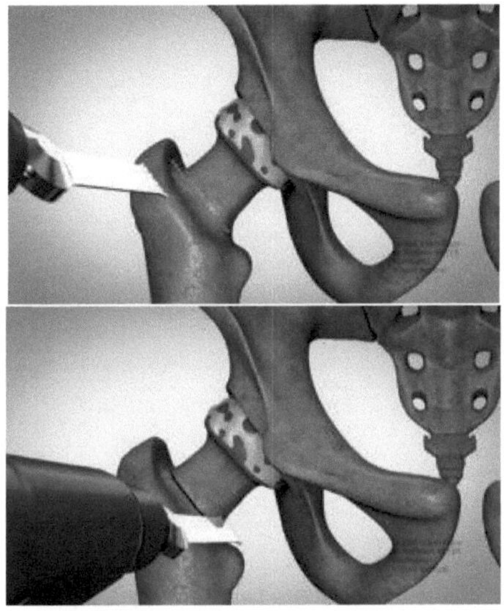

Figure A1- 5: Cross-section of femoral neck

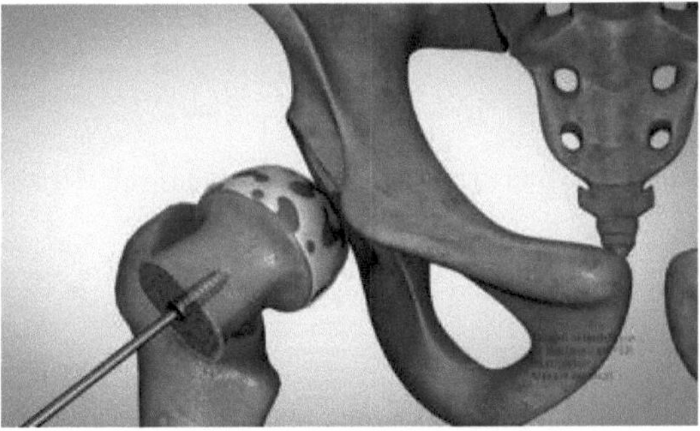

Figure A1- 6: Extraction of the femoral head

The articular cavity of the pelvis is prepared with rotary burs, then the definitive prosthetic cup is impacted. An insert is placed inside this cup, which will articulate with the femoral implant.

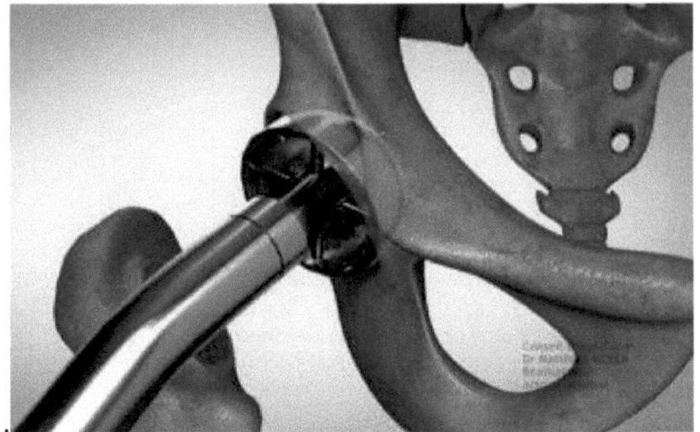

Figure A1- 7: Preparation of the acetabular cavity

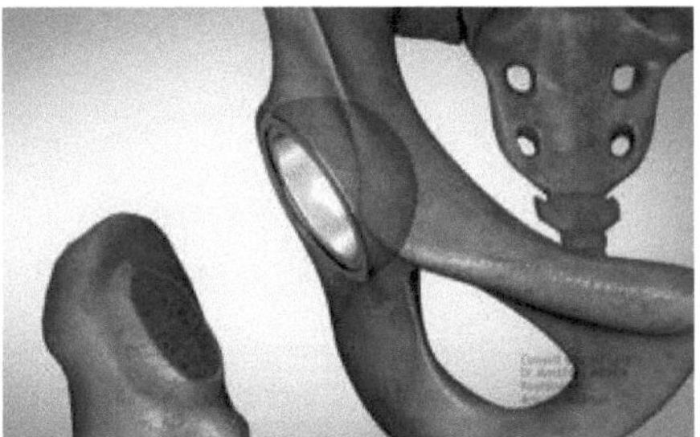

Figure A1- 8: Placement of the acetabular implant

The femur is then prepared to receive the implant: it is catheterized and prepared using rasps of increasing size.

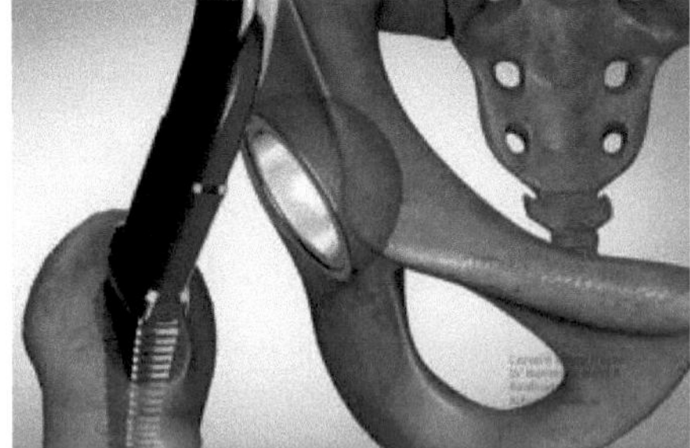

Figure A1- 9: Preparing to receive the femoral stem using rasps

Once the femoral preparation is complete, the definitive prosthetic stem is inserted, onto which the ball that will articulate with the pelvic implant is impacted.

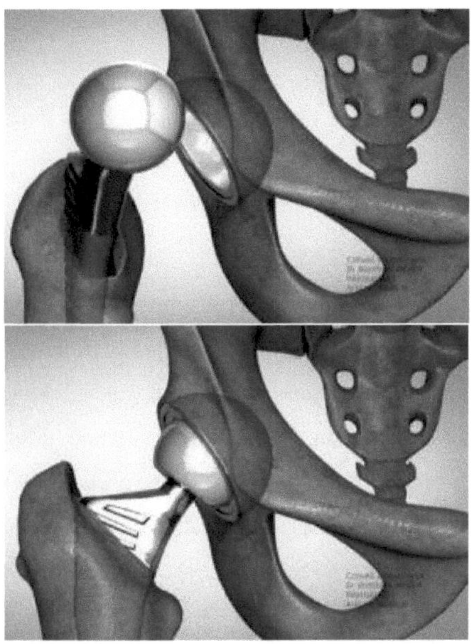

Figure A1- 10: Placing the definitive femoral implant

Once the prosthesis is in place, the capsule can be carefully closed and the retractors removed. The muscles then return to their original positions without having been cut in any way.

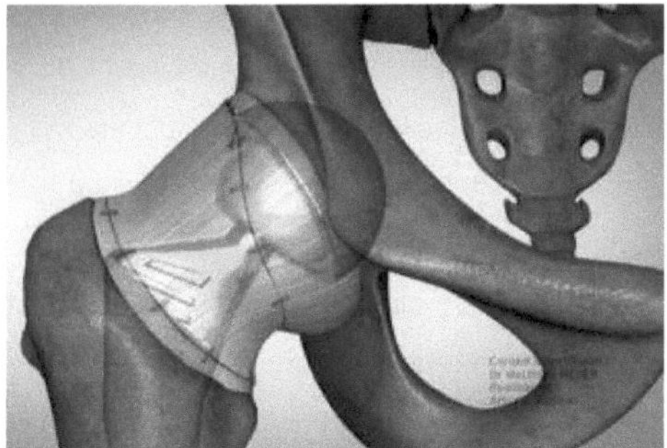

Figure A1- 11: Capsule closure

This preservation of tendon muscles is the main advantage of this surgical technique over other approaches, notably the posterolateral approach still widely used today, which passes through the fibers of the gluteus maximus muscle and cuts the pelvic tendons, destabilizing the joint and increasing the risk of prosthesis dislocation.

A1.2 Complications

Unfortunately, there is no such thing as zero risk in surgery. Every intervention carries risks and has its limits.

Certain risks are common to all types of surgery. Such is the case of infection, where microbes attack the operated area. Fortunately, this is a rare complication, but when it does occur, the prosthesis needs to be washed during a new operation, and antibiotics need to be taken. More rarely, certain infections may require replacement of the prosthesis. Hematomas can also occur in the operated area. This is usually avoided or limited by inserting a suction drain at the end of the procedure, which is removed in the days following the operation. Occasionally, heavy bleeding may require a blood transfusion. In exceptional cases, an operation may be required to evacuate a large hematoma under tension.

Hip surgery also increases the risk of phlebitis, which can be complicated by pulmonary embolism. To minimize this risk, blood-thinning anticoagulant treatment (in the form of daily injections or tablets) is prescribed for the month following surgery.

There are also risks specific to hip replacement surgery. Firstly, the prosthesis may dislocate (dislocation). Dislocation most often occurs in the first few weeks after the prosthesis has been fitted, when everything around it has not yet healed. When the prosthesis is dislocated, a short anaesthetic is required to re-emboss it. It also happens that the two legs are not exactly the same length after the operation. This inequality in length is often well tolerated and goes unnoticed. If this is not the case, and lameness is present, we may prescribe the use of a compensatory sole.

Finally, rarer complications can also occur. A fracture of the femur may occur when it is manipulated during the operation. This usually results in a postponement of the resumption of weight-bearing. Nerves may also be accidentally damaged during the operation, with the risk of paralysis or loss of sensitivity in the operated limb, which may be transient or permanent. [45]

ANNEXES 2

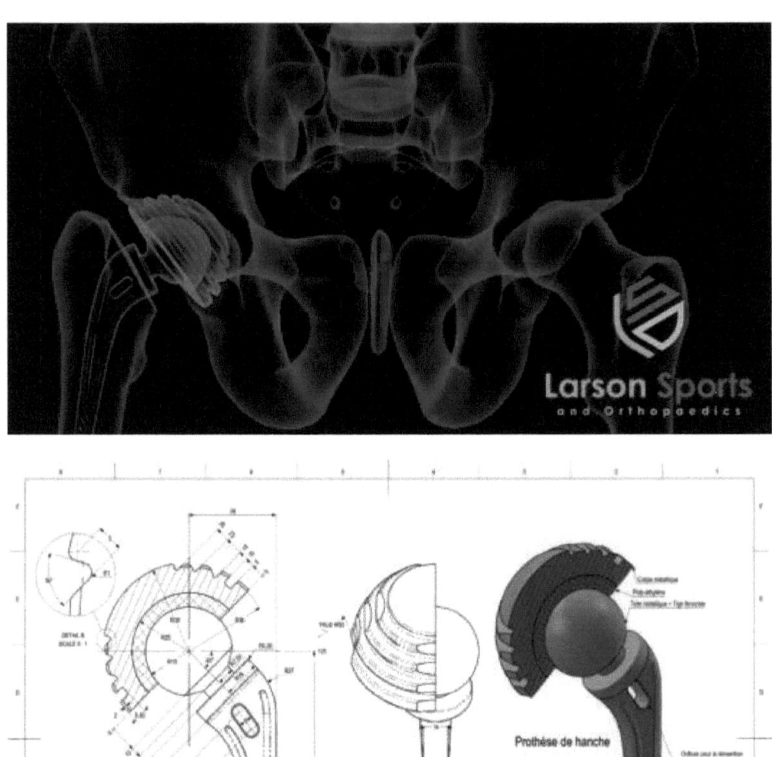

Figure A2- 1: Overall drawing of an actual Larson Sport orthopaedics hip prosthesis model

BIBLIOGRAPHY

[1] Christophe Chevillotte. *Biomechanical Study of Ceramic-Ceramic Friction Torque in Cementless Total Hip Prostheses.* Biomedical engineering. PhD thesis, UNIVERSITE DE LYON, 2012

[2] Armelle Perrichon. *Tribology and aging of bioceramic total hip prostheses, in vitro = in vivo? Scientific and societal issues.* Other. PhD thesis, University of Lyon, 2017.

[3] Juliana Uribe Perez. *Multiscale durability analysis of bioceramics for hip prostheses. In vitro and ex vivo studies.* Other. PhD thesis, Ecole Nationale Supérieure des Mines de Saint-Etienne, 2012.

[4] Jabri, Mariem. *La pose d'une prothèse totale de hanche : contribution de pharmacien hospitalier.* PhD thesis, 2009.

[5] Prigent, François. *The history of hip prostheses.* PhD thesis, 1985.

[6] Fischer, Louis-Paul, Wilfrid Planchamp, Bénédicte Fischer, and Frédéric Chauvin. "Les premières prothèses articulaires de la hanche chez l'homme (1890-1960)." *History of Medical Sciences* 34, no. 1 (2000): 57-70.

[7] "The History of Uncemented Femoral Stems." GILES Group, April 17, 2022. https://groupegiles.org/en/history-femoral-uncemented-femoral-stems/ .

[8] Louboutin, Lucie, Romain Desmarchelier, and Michel-Henry Fessy. "Analyse Du Risque de Descellement Aseptique de La Tige Corail de Troisième Génération (Depuy) à 12ans De Recul." *Revue de Chirurgie Orthopédique et Traumatologique* 102, no. 7 (November 2016). DOI: https://doi.org/10.1016/j.rcot.2016.08.043 .

[9] Puch, Jean-Marc, Guy Derhi, Loys Descamps, Régis Verdier, and Jacques H. Caton. "Dual-Mobility Cup in Total Hip Arthroplasty in Patients Less than Fifty-Five Years and over Ten Years of Follow-Up." *International Orthopaedics* 41, no. 3 (November 8, 2016): 475-80. DOI: https://doi.org/10.1007/s00264-016-3325-x .

[10] Ligia Miriam Amavizca Ruiz. *3D reconstruction of the human pelvis from incomplete multimodal medical images. Application to the assistance of total hip*

replacement (THR) surgery. Automatics / Robotics. PhD thesis, Institut National Polytechnique de Grenoble- INPG, 2005.

[11] Servagent, Raphaël. "3D Planning of a Total Hip Prosthesis." Elsan, April 11, 2014. https://www.elsan.care/fr/nos-actualites/planification-3d-dune-prothese-totale-de-hanche/ .

[12] Mainard, D., O. Barbier, Y. Knafo, R. Belleville, L. Mainard-Simard, and J.-B. Gross. "Accuracy and Reproducibility of La Planification 3d Des Prothèses Totales de Hanche (PTH) Basée Sur La Radiographie Par Balayage Linéaire Biplan à Faible Radiation: Étude Pilote." *Revue de Chirurgie Orthopédique et Traumatologique* 103, no. 4 (June 2017): 377-82. DOI: https://doi.org/10.1016/j.rcot.2017.03.006 .

[13] Odaira, Takumi, Sheng Xu, Kenji Hirata, Xiao Xu, Toshihiro Omori, Kosuke Ueki, Kyosuke Ueda, et al. "Flexible and Tough Superelastic Co-Cr Alloys for Biomedical Applications." *Advanced Materials* 34, no. 27 (June 3, 2022). DOI: https://doi.org/10.1002/adma.202202305 .

[14] Que Choisir. "Prothèses de Hanche: Les Plus Récentes Ne Sont Pas Forcément Les Meilleures." May 30, 2022. https://www.quechoisir.org/actualite-protheses-de-hanche-les-plus-recentes-ne-sont-pas-forcement-les-meilleures-n72451/.

[15] Marwa Ben Braham. *Fatigue and wear behavior of bioceramics used in the design of osteoarticular prostheses.* Biomechanics. PhD thesis, University of Lyon; National Engineering School of Tunis (Tunisia), 2021

[16] Ikrame BERRAMOU. *Total hip prosthesis treatment in hip dysplasia.* PhD thesis, Faculty of Medicine and Pharmacy, Marrakech, 2019

[17] Kharmanda, Ghias, and Abdelkhalak El Ham. *Reliability in Biomechanics: Analysis and Applications.* London: ISTE Editions, 2017.

[18] "Joint prostheses." *Rheumatism.* May 24, 2022. https://www.rhumatismes.net/index.php?id_q=425.

[19] Knee Hip. "Dual Mobility PTH: A French Revolution." Last modified March 18, 2022. https://www.genouhanche.fr/fr/hanche/pth-a-double-mobilite-une-revolution-francaise.html.

[20] Boulila, Atef, Khemaïes Jendoubi, Ali Zghal, Mustapha Khadhraoui, and Patrick Chabrand. "Comportement Mécanique Des Prothèses Totales de Hanche Au Pic de Chargement." *Mécanique & Industries* 11, no. 1 (January 2010): 25-36. DOI: https://doi.org/10.1051/meca/2010013 .

[21] Delikanli, Yunus E, and Mehmet C Kayacan. "Design, Manufacture, and Fatigue Analysis of Lightweight Hip Implants." *Journal of Applied Biomaterials & Functional Materials* 17, no. 2 (April 2019): 228080001983683. DOI: https://doi.org/10.1177/2280800019836830 .

[22] Jerbi, Hana, Daniel Nélias, and Marie-Christine Baietto. "Etude et modélisation de l'endommagement du contact revêtu soumis à des sollicitations de fretting-fatigue." from *Congrès Français de Mécanique (CFM 2015)*, Lyon, France, August 2015.

[23] Mounir Fruja, Tarek Hassine, R. Fathallah, Abdelwaheb Dogui. "Finite Element Modelling of Shot Peening Process: Prediction of the Compressive Residual Stresses, the Plastic Deformations and the Surface Integrity." *Materials Science and Engineering: A* 426, no. 1-2 (June 2006): 173-80. DOI: https://doi.org/10.1016/j.msea.2006.03.097 .

[24] Aid, A. *Cumulative damage in multiaxial fatigue under variable loads*. PhD thesis, University of Sidi-Bel Abbes, Algeria, 2006.

[25] Migaud, Henri, Julien Girard, Olivier May, Marc Soenen, Yannick Pinoit, Philippe Laffargue, and Gilles Pasquier. "Les Arthroplasties de Hanche Aujourd'Hui: Principaux Matériaux, Voies d'abord." *Revue du Rhumatisme* 76, no. 4 (April 2009): 367-73. DOI: https://doi.org/10.1016/j.rhum.2008.04.026 .

[26] Subhedar, Prajakta, Gajanan Thokal, and C.R. Patil. "Carbon Fiber Reinforced Polyamide12 as a Biomaterial for Implant." *International Journal of Current Engineering and Technology* 6, no. 2 (April 2016): 568-571.

[27] CTIF. "Titanium Alloys for Medical Applications." March 5, 2018. https://metalblog.ctif.com/les-alliages-de-titane-pour-le-medical/.

[28] Kumar, Abhinav, Apoorv Rathi, Jagjit Singh, and N. K. Sharma. "Studies on Titanium Hip Joint Implants Using Finite Element Simulation." In *Proceedings of*

the World Congress on Engineering, Vol. 2, 986-990. London, U.K.: IAENG, 2016. DOI: http://dx.doi.org/10.13140/RG.2.1.3851.0325 .

[29] Colic, Katarina, Aleksandar Sedmak, Aleksandar Grbovic, Uros Tatic, Simon Sedmak, and Branislav Djordjevic. "Finite Element Modeling of Hip Implant Static Loading." *Procedia Engineering* 149 (2016): 257-62. DOI: https://doi.org/10.1016/j.proeng.2016.06.664 .

[30] MatWeb. "Material Property Data." Last modified May 2, 2022. https://www.matweb.com.

[31] Chahid, Younes. "How to Design and Optimize a Patient Specific Additively Manufactured Hip Implant Stem." *nTop*, April 27, 2020. https://www.ntop.com/resources/blog/how-to-design-and-optimize-a-patient-specific-additively-manufactured-hip-implant-stem/.

[32] AMFG. "Application Spotlight: 3D Printing for Medical Implants." March 22, 2022. https://amfg.ai/application-spotlight-3d-printing-for-medical-implants/.

[33] I. Eldesouky, O. Abdelaal and H. El-Hofy, "Femoral hip stem with additively manufactured cellular structures," *2014 IEEE Conference on Biomedical Engineering and Sciences (IECBES)*, Kuala Lumpur, Malaysia, 2014, pp. 181-186, DOI: https://doi.org/10.1109/IECBES.2014.7047482 .

[34] Annanto, Gilar Pandu, Eko Saputra, J. Jamari, Athanasius Priharyoto Bayuseno, Rifky Ismail, Mohammad Tauviqirrahman, and Iwan Budiwan Anwar. "Numerical Analysis of Stress Distribution on Artificial Hip Joint Due to Jump Activity." In *Proceedings of the E3S Web of Conferences,* 73(), 12005-. DOI: https://doi.org/10.1051/e3sconf/20187312005 .

[35] AMTI. "Hip Simulator." June, 2022. https://www.amti.biz/product/hip-simulator/.

[36] Galanis, Nikolaos I., and Dimitrios E. Manolakos. "Design of a Hip Joint Simulator According to the ISO 14242." In *Proceedings of the World Congress on Engineering*, Vol. III, 2088-2093. London, U.K., July 6-8, 2011.

[37] Kedadria, Abderrazak. 2018. "Hip Simulator." August 14. https://grabcad.com/library/hip-simulator-1.

[38] De Pieri, E., D.E. Lunn, G.J. Chapman, K.P. Rasmussen, S.J. Ferguson, and A.C. Redmond. "Patient Characteristics Affect Hip Contact Forces during Gait." *Osteoarthritis and Cartilage* 27, no. 6 (June 2019): 895-905. DOI: https://doi.org/10.1016/j.joca.2019.01.016.

[39] Bergmann, G, G Deuretzbacher, M Heller, F Graichen, A Rohlmann, J Strauss, and G.N Duda. "Hip Contact Forces and Gait Patterns from Routine Activities." *Journal of Biomechanics* 34, no. 7 (July 2001): 859-71. DOI: https://doi.org/10.1016/s0021-9290(01)00040-9 .

[40] Bergmann, G., F. Graichen, and A. Rohlmann. "Hip Joint Loading during Walking and Running, Measured in Two Patients." *Journal of Biomechanics* 26, no. 8 (August 1993): 969-90. DOI: https://doi.org/10.1016/0021-9290(93)90058-m .

[41] Taleb Hosni Abderrahmane. *Chapter 3: Laws of Behavior.* Center Universitaire de Mila, January 2021, p45.

[42] Abdulkader Zalt. *Fatigue damage and service life prediction of welded box-type structures.* Other. PhD thesis, University of Lorraine, 2012.

[43] Ritwik Bandyopadhyay. *Ensuring fatigue performance via location-specific lifing in aerospace components made of titanium alloys and nickel-base superalloy.* These de doctorat, Purdue University (2020). DOI: https://doi.org/10.25394/PGS.12168018.v1 .

[44] Imade Koutiri. *Effect of high hydrostatic stresses on the fatigue life of metallic materials.* Mechanics of materials. PhD thesis, École nationale supérieure Arts et Métiers ParisTech, 2011

[45] Meyer, Matthieu. "Mini-Invasive Anterior Route Hip Prosthesis." *Dr. Meyer Orthopédie.* Last modified April 16, 2022. https://www.dr-meyer-orthopedie.fr/operations/hanche/prothese-totale-de-hanche/prothese-hanche-voie-anterieure-mini-invasive.

Printed by Books on Demand GmbH, Norderstedt / Germany